RIP CURRENT PREDICTION S
SWIMMER SAFETY

T0231305

TOWARDS OPERATIONAL FORECASTING USIN
MODEL AND NEARSHORE BATHYMETRY FROM VIDEO

RIP CURRENT PREDICTION SYSTEM FOR SWIMMER SAFETY

TOWARDS OPERATIONAL FORECASTING USING A PROCESS BASED MODEL AND NEARSHORE BATHYMETRY FROM VIDEO

DISSERTATION

Submitted in fulfilment of the requirements of
the Board for Doctorates of Delft University of Technology and of
the Academic Board of the UNESCO-IHE Institute for Water Education
for the Degree of DOCTOR
to be defended in public
on Friday, October 30, 2015, at 10:00 hours
in Delft, The Netherlands

by

Leo Eliasta SEMBIRING

born in Kabanjahe, Indonesia
Bachelor of Engineering in Civil Engineering, Bandung Institute of Technology
Master of Science in Coastal Engineering, UNESCO-IHE, Delft

This dissertation has been approved by the promotor:
Prof. dr. ir. J. A. Roelvink

Composition of the Doctoral Committee:

Chairman	Rector Magnificus, Delft University of Technology
Vice-Chairman	Rector UNESCO-IHE
Prof. dr. ir. J.A. Roelvink	UNESCO-IHE/ Delft University of Technology,
promotor	

Independent members:

Prof. dr. R.A. Holman	Oregon State University, USA
Dr. M. J. Austin	Bangor University, UK
Prof. dr. ir. M.J.F. Stive	Delft University of Technology
Prof. dr. R. Ranasinghe	Australian National University, Australia/UNESCO-IHE
Dr. A.R. van Dongeren	Deltares, Delft, the Netherlands
Prof. dr. ir. A.E. Mynett	Delft University of Technology, reserve member

CRC Press/Balkema is an imprint of the Taylor & Francis Group, and informa business

© 2015, Leo Eliasta Sembiring

All rights reserved. No part of this publication or the information contained herein may be reproduced, stored in a retrieval system, or transmitted in any form or by any means, electronic, mechanical, by photocopying, recording or otherwise, without written prior permission from the publishers.

Although all care is taken to ensure the integrity and quality of this publication and information herein, no responsibility is assumed by the publishers or the author for any damage to property or persons as a result of the operation or use of this publication and or the information contained herein.

Published by:
CRC Press/Balkema
PO Box 11320, 2301 EH Leiden, the Netherlands
Email: pub.nl@taylorandfrancis.co.uk
www.crcpress.com – www.taylorandfrancis.com

ISBN 978-1-138-02940-8 (Taylor & Francis Group)

This research has been supported by:

Summary

Rip currents are among the most dangerous coastal hazards for the bathing public, and contribute to the highest portion of beach rescues all over the world. In order to help life guards in planning and preparing rescue resources so as to minimize casualties, information about where and when rip currents may occur is needed. This can be provided by a predictive tool which combines meteorological forecasts, hydrodynamic models and remote-sensed observations. However, to implement this approach for the nearshore at a beach resort, up-to-date and high resolution bathymetry data are needed since the time scale of nearshore morphology change can be in the order of days to weeks, depending on the environmental (waves, hydrodynamic) conditions. Therefore, having accurate bathymetry data based on conventional bathymetry surveys on this time scale would require a large logistical effort, and would be very costly.

The objective of the current research is to develop and test a methodology with which forecasts of rip currents can be provided for swimmer safety purposes at Egmond aan Zee. An operational model system, CoSMoS (Coastal Storm Modelling System), developed by Deltares, will be used as the main task manager to control the operation of numerical models. A validation of CoSMoS will be carried out to evaluate the performance of the model system in providing waves and water level boundary conditions. Next, to evaluate the model's ability in simulating rip currents at Egmond aan Zee, numerical experiments simulating rip currents are set up and results are validated using data from field experiments. Furthermore, the numerical experiments will be extended, and a longer period is simulated in order to gain more knowledge on rip current characteristics at Egmond aan Zee and the safety implications. To meet the need of continuous and up to date bathymetry, a technology using video images to predict nearshore bathymetry on a daily basis, cBathy, will be utilized.

Validation of the CoSMoS model system shows that the CoSMoS Dutch Continental Shelf Model is a fit-for-purpose regional model to simulate waves, tide and surge, including the interactions, in order to provide boundary conditions for coastal models. Hindcast results over the whole year of 2009 show that the simulated wave parameters and surge elevation from the CoSMoS are in good agreement with the data. It is noted that there is a tendency of the wave model to underestimate the height of northerly waves with lower frequencies (swell). Additionally, when a wave separation algorithm is applied to the overall spectrum, results show consistent underestimation of the swell component by the model, which for the Dutch coast will mainly come from the north, where the North Sea is open to the Atlantic Ocean. In the proposed model system, the swell boundary can have a significant effect on the simulated wave results, suggesting room for improvement for the swell boundary conditions to the north and the swell propagation within the Dutch Continental Shelf Model. Furthermore, when the CoSMoS was run in forecast mode, it can provide reasonably good wave and surge prediction.

The next part of the thesis presents rip current modelling. From the numerical experiments, it is found that wave height and water level strongly control the initiation and the duration of the rip currents at Egmond. For the period of analysis, the rip currents are initiated approximately 5 hours before low tide, reach their peak during (peak) low tide, start to decay as the tide is rising, and finally become inactive 3 hours after low tide. Their initiation corresponds to the ratio of offshore wave height to water depth on the bar of ~0.55. Rips may also occur during the high tide (when the 0.55 ratio is fulfilled), which requires a relatively high wave height. In addition, it also found that within 5 minutes, an object can be transported as far as 60 meters offshore from the rip channel, even 4 hours before the low tide. As the tide is approaching the low tide, this number increases to ~90 m during the (peak) low tide. This finding defines the threat of the rip at Egmond.

Nearshore bathymetry can be accurately obtained from video images using the cBathy technique. cBathy performs very well in predicting morphological features both in alongshore and cross shore orientation, and is designed to provide daily estimates. In the very shallow areas near the shoreline, cBathy performs poorly, which can be mitigated by integrating cBathy estimates with intertidal bathymetry obtained from shoreline detection technique. The integration significantly improves the bathymetry estimates in these regions. Moreover, cBathy does not require prior image quality selection, and is designed to provide bathymetry estimates a few times per day. Image collection together with the analysis process is performed fully automatically. Therefore, from an operational point of view, cBathy shows a great potential to be applied.

Applying the video bathymetry in numerical modelling shows that nearshore currents simulated using video bathy agree very well with those using surveyed/ground truth bathymetry. Coupling the video bathymetry estimates with CoSMoS in forecast mode shows that dangerous rips were predicted well, which was verified using rip incident reports posted by the lifeguards through their Twitter page. This confirms the potential application of the proposed system in providing forecasts for rip currents at Egmond aan Zee.

Samenvatting[1]

Muistromen behoren tot de meest gevaarlijke kust-risico's voor het badpubliek en dragen in de hele wereld bij aan het grootste deel van de strand-reddingsacties. Teneinde strandwachten te helpen beter hun personeel en middelen te plannen om het aantal slachtoffers te minimaliseren is informatie nodig over waar en wanneer muistromen te verwachten zijn. Hierin kan worden voorzien door een voorspellend model waarin meteorologische voorspellingen, hydrodynamische modellen en remote-sensing observaties worden gecombineerd. Echter, om deze aanpak voor de nabije kustzone in een badplaats te implementeren is actuele en hoge-resolutie bathymetrie-data nodig, aangezien de tijdschaal van kustnabije bodemveranderingen in de orde van dagen tot weken kan zijn, afhankelijk van de omgevingscondities (golven, stromingen). Daarom zou het uitvoeren van conventionele dieptemetingen op deze tijdschaal een grote logistieke inspanning vergen en erg kostbaar zijn.

Het doel van deze studie is, een methodologie te ontwikkelen en te testen waarmee voorspellingen van muistromen kunnen worden gemaakt voor zwemveiligheids-doeleinden, bij Egmond aan Zee. Een operationeel modelsysteem, CoSMoS (Coastal Storm Modelling System), ontwikkeld door Deltares, zal worden gebruikt om het draaien van de numerieke modellen te coordineren. Een validatie van CoSMoS zal worden uitgevoerd om de performance van het modelsysteem in het leveren van golf- en waterstands-randvoorwaarden te testen.

Vervolgens, om te evalueren of het model de muistromen bij Egmond aan Zee kan simuleren, worden numerieke experimenten opgezet en de resultaten gevalideerd aan de hand van data uit veldexperimenten. Daarna worden de numerieke experimenten uitgebreid en wordt een langere periode gesimuleerd om meer inzicht te krijgen in mui-karakteristieken bij Egmond aan Zee en in de veiligheids-implicaties daarvan. Om tegemoet te komen aan de behoefte aan continue en actuele bodemdiepte-informatie wordt een technologie op basis van video-waarnemingen, cBathy, toegepast.

Validatie van het CoSMoS modelsysteem laat zien dat het CoSMoS Dutch Continental Shelf model een geschikt regionaal model is om golven, getij en stormopzet te simuleren, inclusief hun interacties, om randvoorwaarden te genereren voor kustmodellen. Hindcast resultaten over het gehele jaar 2009 laten zien dat de gesimuleerde golfparameters en stormopzet uit CoSMoS in goede overeenstemming zijn met de metingen. Er is een tendens in het golfmodel om de hoogte van noordelijke golven met lage frequenties (deining) te onderschatten. Bovendien, wanneer een golf-separatie algoritme wordt toegepast op het gehele spectrum, laten de resultaten een consistente onderschatting van de deiningscomponent door het model zien, die voor de Hollandse kust voornamelijk uit het noorden zal komen, waar de Noordzee open is naar de Atlantische Oceaan. In het voorgestelde modelsysteem kan de deiningsrandvoorwaarde een significant effect hebben op de gesimuleerde golfresultaten, hetgeen ruimte voor verbetering suggereert in de deiningsrandvoorwaarden aan de noordzijde en de voortplanting van de deining in het

---

[1] This summary is translated into Dutch by Prof. J.A. Roelvink

Dutch Continental Shelf Model. Verder, wanneer CoSMoS in voorspellings-mode wordt gedraaid, kan het redelijk goed de golf en stormopzet weergeven.

Het daarop volgende deel van het proefschrift behandelt de muistromings-modellering. Uit de numerieke experimenten wordt gevonden dat golfhoogte en waterstand sterk de initiatie en duur van de muistromingen bij Egmond bepalen. Voor de periode van de analyse geldt dat de muistromingen ongeveer 5 uur voor laagwater beginnen, hun piek bereiken tijdens laagwater, beginnen af te nemen bij rijzend tij en uiteindelijk 3 uur na laagwater inactief worden. Hun initiatie valt samen met een verhouding tussen zeewaartse golfhoogte en waterdiepte op de bank van ~0.55. Muien kunnen ook voorkomen gedurende hoogwater (wanneer de 0.55 verhouding wordt gehaald), hetgeen een relatief grote golfhoogte vereist. Bovendien wordt gevonden dat een object binnen 5 minuten tot wel 60 m zeewaarts kan worden getransporteerd, zelfs 4 uur voor laagwater. Als het getij laagwater nadert kan dit oplopen tot ca. 90 m. Deze bevinding maakt het gevaar van de mui bij Egmond duidelijk.

Kustnabije bathymetrie kan nauwkeurig worden geschat uit videobeelden met behulp van de cBathy techniek. cBathy is heel goed in het voorspellen van morfologische kenmerken, zowel in kustlangse als in kustdwarse richting, en is ontworpen om dagelijkse schattingen te leveren. In de heel ondiepe zones bij de waterlijn doet cBathy het matig, het geen kan worden opgevangen door cBathy schattingen te integreren met intergetijde-bathymetrie verkregen met een waterlijn-detectie techniek. De integratie verbetert de bathymetrie schattingen in deze gebieden significant. Bovendien heeft cBathy geen a priori beeldkwaliteitselectie nodig en is het ontworpen om een paar keer per dag bathymetrie schattingen te leveren. Beeldinwinning tezamen met het analyseproces wordt volledig automatisch uitgevoerd. Daarom biedt cBathy vanuit een operationeel gezichtspunt een grote potentie om te worden toegepast.

Toepassing van de video bathymetrie in numerieke modellering laat zien dat kustnabije stromingen gesimuleerd op basis van bathymetrie uit video goed overeenkomen met die op basis van conventioneel gemeten bathymetrie. Koppelen van de video bathymetrie met CoSMoS in voorspellings-mode laat zien dat gevaarlijke muistromingen goed voorspeld werden, hetgeen is geverifiëerd op basis van de 'incident reports' gepubliceerd door de strandwachten op hun Twitter pagina. Dit bevestigt de potentiële toepassing van het voorgestelde systeem in het voorspellen van muistromen bij Egmond aan Zee.

Contents

1 Introduction

1.1 Problem statement

Rip currents are among the most dangerous coastal hazards, and account for the highest portion of beach rescues (Lushine, 1991; Lascody, 1998; Short, 2007). In East Florida, there is an average of 21 drownings per year, which is larger than the number of deaths due to tornados, thunderstorms, lightning, and hurricanes combined. In the UK, data from the Royal National Lifeboat Institution (RNLI) show that 71% of all recorded incidents taken from 62 beaches in the southwest of England were due to rip currents (Scott *et al.*, 2007). In Australia, rip currents are recognized as the major hazards to beach going public, as they are responsible for more than 90% of all beach rescues (Short, 1999). Many other countries such as Israel, Brazil, and Colombia have also reported that rip currents are a major hazard (see Short, 2007).

In the Netherlands, the numbers are not as large as those reported above. However, the numbers of drownings is increasing (based on direct interviews with local lifeguards at Egmond, the Netherlands). The victims are mostly school children, but adults have also been reported as victims. Warnings are provided through a flag system (for instance a yellow flag means swimming is allowed but not floating objects, and a red flag means swimming is prohibited), and published through the lifeguard's website. The warnings provided are based on practical experience of the lifeguards. Dangerous conditions are categorized mainly based on wave heights.

The need of a predictive tool comes from the lifeguards on site, which identifies a necessity to minimize the risk. Information about where and when rip currents are likely to occur during the next day and when dangerous situations exist is eagerly awaited. To provide lifeguards with such information, an approach that can be implemented is to make use of computer process-based (i.e., based on physics) models as tools, to simulate and predict the occurrence of rip currents. Process based models have been widely used for various purposes, starting from the simplistic 1D approach for river and coastal applications to complex 3D simulations of estuary systems (e.g. see Lesser *et al.*, 2004; Roelvink *et al.*, 2009; Larsen *et al.*, 2013). The application of this process-based model approach has been reported to be successful in providing us information on the main physical interactions of the system to be understood. However, to implement this approach to the nearshore area of a beach resort, up-to-date and high resolution bathymetry data are needed. These data can be obtained by remote-sensing techniques at relatively low costs, and can be integrated in an operational system, as will be shown in this thesis.

1.2 Objectives and research approach

1.2.1 Objectives

The objective of the present study is to develop and test a methodology with which forecasts of rip currents can be provided for swimmer safety purposes.

1.2.2 Research questions

The research questions are:

1. Can we predict the occurrence, duration, and magnitude of the rip currents at Egmond using process-based models? What is the added-value for swimmer safety warning systems?
2. Can we obtain nearshore bathymetries through video techniques for Egmond aan Zee in an operational mode?
3. Can we apply nearshore bathymetry from video to predict nearshore currents and rip currents?

1.3 Approach

1.3.1 Coastal operational model

As the main task manager of the proposed system, the operational model system CoSMoS (Coastal Storm Modelling System, Van Ormondt *et al.*, 2012) will be used. This operational model system combines different models with different spatial scales in a "nesting" manner. By "nesting", larger-scale models provide boundary conditions to smaller-scale but higher-resolution models. A validation of CoSMoS will be carried out to evaluate the performance of the model system in predicting waves and water level in deep water.

1.3.2 Rip currents numerical modelling

To address Research Question number 1, numerical experiments have been performed. During the study, a field campaign was conducted on August 2011, where bathymetry data at Egmond aan Zee beach were collected. In addition, Lagrangian mean currents in the vicinity of the rip channels were also measured by deploying GPS tracked drifters. These data will be used to validate the rip resolving model built for Egmond aan Zee, in which dynamics due to the tidal currents are taken into account. Further, the numerical experiment has been extended to gain more knowledge on rip currents' initiation and duration at Egmond aan Zee and on the added value for swimmer safety.

1.3.3 Nearshore bathymetry from ARGUS video

Dealing with accurate and up to date bathymetry, a technology using video images has been extensively investigated worldwide, which allows us to utilize bathymetry from the sets of video image data, called, ARGUS (see Holman and Stanley, 2007). At Egmond aan Zee, five ARGUS cameras are installed at the Jan Van Speijk lighthouse, covering approximately 3 kilometers of coastline. This video technique was first initiated by the Coastal Imaging Lab (CIL) at Oregon State University under the Coast View Program. Physically, an Argus Station consists of a number of video cameras attached to a host computer that serves as both system control and a communication link between the cameras and central data archive. A standard collection scheme normally consists of several types of images: a snapshot image, ten minutes time exposure images, variance images, and daily time exposure images. Depending on the method being used, these images can be utilized to obtain nearshore bathymetry. Several ARGUS application works have been conducted previously e.g. for intertidal mapping (Plant and Holman, 1997; Aarninkhof *et al.*, 2003; Uunk *et al.*, 2010), and sub-tidal bathymetry (Stockdon and Holman, 2000; Aarninkhof *et al.*, 2005). In addition, a data assimilation technique has also been introduced to estimate nearshore bathymetry (e.g. Scott and Mason, 2007; van Dongeren *et al.*, 2008; Wilson *et al.*, 2010). Recently, a method called cBathy (Holman *et al.*, 2013) has been developed, which is able to provide 2 dimensional bathymetry estimates on a daily basis with a fully automated process, showing a great potential to be applied as a forecasting system component like the one proposed in this study. Applying cBathy on wind sea dominated environments to provide bathymetry data and use it for rip current predictions has not been demonstrated yet. This will address Research Question number 2.

1.3.4 Prediction system

In this thesis, a methodology will be developed where we combine the CoSMoS model system as a boundary generator to a beach scale numerical model, and the bathymetry boundary will be obtained from video technique (see Figure 1-1). In order to test the proposed system, firstly, applicability of nearshore bathymetry obtained from video images will be evaluated by using numerical experiments where nearshore currents simulated by models using video bathymetry are compared with ones using ground truth bathymetry. Afterwards, a test case will be performed in which the CoSMoS model system will be initiated in forecast mode in order to obtain quasi-operational simulation results for Egmond aan Zee. The test case will be for the summer period of the year 2013, following the availability of video data to be used by cBathy algorithm described in the previous section. The forecast results will be qualitatively verified using rip incident data obtained from the Twitter[©2] Page of the Life Guard organization at Egmond aan Zee. This will address Research Question number 3.

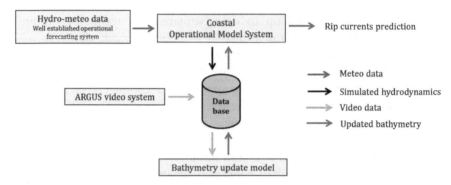

Figure 1-1: Conceptual diagram of the proposed rip current prediction system

1.3.5   Case study site

The case study site, Egmond aan Zee, is a busy beach resort during the summer season, located in the middle of the uninterrupted part of the Dutch Coast between the IJmuiden (port of Amsterdam) harbour moles and the Wadden Sea on the north. It is characterized by a double sand bar with reset events for the outer bar on the time scale of ~5 years (Walstra *et al.*, 2012). Wave conditions are mild with 1.3 m mean significant wave height and 6.4 seconds wave period respectively (Wijnberg, 2002), and significant seasonal variations between summer and winter in which during winter waves are higher. The beach is meso-tidal, with a tidal range of ~ 1.4 m – 1.7 m (Wiersma and Van Alphen, 1987). The tidal current is a pronounced hydrodynamic feature at this site, where strong alongshore (tidal) velocity flows to the north during flood tide and to the south during ebb tide. The rip channels interrupt the sand bar, which are located approximately 70-100 meters from the mean shoreline. The depth of the rip channels is on average 1.5 meters. On the beach, rip currents occur which are

---

[2] Twitter, is a free service of an online social networking that enables registered user to send and read short messages usually called 'tweets' (https://twitter.com/)

induced by the underlying bathymetry (a term *bathymetrically controlled-rip currents* will be used from here on).

1.4 Outline

The thesis is opened with the general introduction and problem statement in Chapter 1. Also in this first chapter, the research questions to be answered and the approach are presented.

In Chapter 2, a literature review on bathymetrically controlled rip currents' physics and generation mechanisms is presented. In addition, reviews on rip current prediction system developed worldwide, including some recent developments are summarized. These include both data driven-prediction systems and physics-based prediction systems.

In Chapter 3, the operational model CoSMoS is presented. Here, the set-up of this model system will be elaborated, and followed by validation of the coupled wave-tide-surge models within the system using field data from sites along the Dutch Coast.

The timing of rip currents is examined using numerical experiments. In Chapter 4, these numerical experiments are presented. The model will be validated using Lagrangian current measurements collected from the vicinity of the rip channels. Further, the model will be used to examine the initiation and duration of rip currents at Egmond aan Zee. Moreover, the implications for swimmer safety will also be analysed based on numerical experiment results.

In Chapter 5, a study on obtaining beach bathymetry from video techniques is presented. The chapter will be opened with a brief review on nearshore bathymetry estimation through remote sensing technology, especially video techniques. Afterwards, two techniques: Beach Wizard and cBathy will be discussed, in which theoretical background, field application, and analysis on the potential application in the operational mode of each technique will be presented. In addition, integration of sub-tidal bathymetry estimates and intertidal bathymetry from shoreline detection technique, in order to obtain a comprehensive beach bathymetry, will also be presented in this chapter.

Chapter 6 is an 'application case' chapter. Here, simulations using video bathymetry are compared with ones using ground truth bathymetry. The next part of this chapter will present a test case of rip current forecasting. The CoSMoS model system will be initiated and forced by forecast hydro-meteorological data, in order to provide quasi-operational simulations. The bathymetry boundary will be obtained from the video estimates. This chapter will demonstrate the potential application of the proposed forecasting system to predict rip currents at Egmond aan Zee.

The present study will be summarized in Chapter 7. An outlook for the application of the proposed forecast system and how the results can usefully be presented to the public and life guard organizations will be presented in this chapter. This will be followed by some potential future research topics.

2 Literature review on rip currents and rip current prediction systems

2.1 Introduction

Rip currents have been a research focus for many scientists in the last decades, since they are very visible and dangerous features in the nearshore zone. There are many publications putting emphasis on kinematics and characteristics of rip currents In contrast, the amount of literature discussing about predicting rip currents as part of swimmer safety approach are not as many as those of rip current theory, although at least 80% of beach rescues and accidents are related to rip currents. In the following, reviews of rip currents and the prediction systems are summarized, partly adopted from Swinkels, 2011.

2.2 Rip current review

2.2.1 Generation and forcing of rip currents

There are in general three kind of rip current generation models as summarized by Bruijn, 2005, based on observation of rip currents in The Netherlands. Each model will be elaborated in the following.

The first model is based on the concept that rip currents can be initiated due to an alongshore variation in wave forcing. This concept, delivered by Bowen and Inman, 1969, is founded on a series of papers on the radiation stress concept delivered by Longuet-Higgins and Stewart, 1962; Longuet-Higgins and Stewart, 1963; Longuet-Higgins and Stewart, 1964. This model simply stated that rip currents exist due to variation of radiation stress component in alongshore direction. According to Bowen and Inman, 1969, alongshore variations of the onshore directed wave force are direct consequences of alongshore variations in wave height. They attribute these alongshore variations to the presence of edge waves (see also Vittori *et al.*, 1999).

The second concept which describes the existence of rip currents can be found in Dalrymple and Lozano, 1978. They stated that rip currents are maintained by wave-current interaction. The wave current-interaction in this approach is due to the refraction of incoming wave towards the rip currents, as waves which are approaching the shoreline will refract towards the rip. This will result in variation in radiation stress (alongshore) component, and generate alongshore wave force towards the origin location of the rip currents. Given by assumption that the rip currents already exist, this system will form a circulation cell which maintain the offshore-ward flow. However, this paper does not really emphasize the genesis of rip currents.

The third approach describing the mechanism of rip current occurrence was delivered by Caballeria *et al.*, 2002. In principle, this explanation is similar to the one delivered by Bowen and Inman, 1969, which stated that variations in alongshore orientation of onshore directed wave force generate offshore-ward flow. In this paper, the alongshore variation is defined explicitly by variations in bottom level. Moreover, they stated that this variation in bottom level is dynamic enough to maintain a regular spacing due to the feedback mechanism between flow and erodible bed, implicitly stating that morphological changes due to the flow control the bottom perturbation.

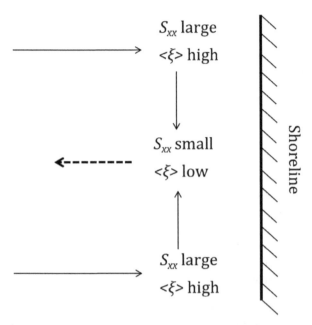

Figure 2-1: Schematized horizontal momentum and wave force flow due to variation in alongshore direction, adapted from Bruijn, 2005

Nowadays, the commonly acceptable model and explanation of rip currents' genesis is the one proposed by Caballeria *et al.*, 2002, since this approach does not need existence of external hydrodynamic forcing (e.g. edge waves). In addition, the model described by Dalrymple and Lozano, 1978 also did not explain explicitly origination of rip currents, instead just emphasizing that rip currents can be maintained by wave-current interaction. Some studies and models now have supported the fact that the generation of rip currents is mainly due to the alongshore variations of the onshore directed wave force.

To summarize, rip currents are generated mainly due to a variation in alongshore direction in wave induced force, or also possibly due to wave height variability in alongshore direction. When waves are propagating shoreward, there is a transport of momentum, well known as radiation stress (Following the definition by Longuet-Higgins and Stewart, 1964, radiation stress S_{ij} is total depth integrated flux of i momentum in j direction). Since wave height decreases due to breaking, a radiation stress gradient develops as wave height decays over x (cross shore) direction, which will induce (wave) force on the water body

$$F_x = -\frac{\partial}{\partial x}(S_{xx})$$

2-1

In Equation 2-1, S_{xx} is the radiation stress (for simplification, consider 1 dimension in cross shore orientation), and F_x is the wave-induced force. This force has to be balanced in order to satisfy momentum balance equation. Therefore, the wave-induced force will be balanced with pressure gradient (water level set up):

$$F_x = \rho g(h+\xi)\frac{\partial \xi}{\partial x}$$

2-2

In Equation 2-2, h is water depth, ξ is water elevation, ρ and g are constant for water density and acceleration due to gravity respectively. The variation of water level set up alongshore, will result in flow convergence toward the rip neck (alongshore arrows in Figure 2-1), in other words, to place where water surface gradient relatively lower, which is the rip current. The variability, again mainly due to variation in underlying bathymetry, but can be in some cases, due to variation of wave height in alongshore direction.

The variations in wave heights alongshore can also be due to refraction of waves which are travelling across offshore submarine canyon (Long and Özkan-Haller, 2005). Rip currents can also be generated by large scale meandering alongshore flow, which can trigger so called "shear instabilities" and generate offshore directed flow, known as transient or flash rips (Özkan-Haller and Kirby, 1999; Johnson and Pattiaratchi, 2004; Murray *et al.*, 2013; Castelle *et al.*, 2014), or due to the presence of structures (e.g. Dalrymple *et al.*, 1977; Pattiaratchi *et al.*, 2009).This type of rip current flow will not be part of the analysis in this study, and focus will be on rip currents that originate from alongshore variability in wave forcing due to alongshore bathymetrical variation.

2.2.2 Bathymetrically controlled rip currents

Morphologically controlled rip currents are known as the most common type of rip currents. The locations of morphologically controlled rip currents are very commonly coupled with the so called rip channels, which normally interrupt sand bar. This channel can be initially a small bottom variation in alongshore direction, but potentially grow rapidly through the feedback mechanism of flow and erodible bottom, forming a distinct rip channel. Beach reset due to storm event can also result in small variation in the alongshore direction as initial state. Using beach definition from Wright and Short, 1984, rip channels can be found on all intermediate beaches. Among those types of beaches, Rhythmic Bar and Beach (RBB) and Transverse Bar and Rip (TBR) type of beaches exhibit the most alongshore variability of bottom morphology (Ranasinghe *et al.*, 2004), thus show distinct rip channel patterns.

2.2.3 Numerical modelling of rip currents

Svendsen *et al.*, 2000 investigated the effect of the rip spacing and the bar distance from the shoreline, using the quasi 3D numerical model SHORECIRC (Putrevu and Svendsen, 1999; Van Dongeren *et al.*, 1994). They concluded that in principle, rip currents are local flow patterns, meaning that a certain rip cell is not influenced by neighbouring rips if their alongshore spacing is larger than 4-8 times the width of the rip channel. This will also mean that the rip strength is independent of rip spacing. They also found that in a certain rip cell, the system behind the bar plays an important role in the volume exchange.

Haller *et al.*, 2002 presented measured waves and currents from a set of laboratory experiments, setting up a fixed barred bathymetry accompanied with periodically spaced rip channels. The laboratory model was forced by a range of incident wave conditions. They found that, when a rip current is strong enough, current-induced breaking can take place which can lead to high wave dissipation, which means a strong rip can weaken the radiation stress gradients, therefore forces opposing the feeder channel significantly reduce. This will result in stronger feeder and thus stronger rip. This research suggests considering this positive feedback mechanism for future modelling of rip currents.

Chen *et al.*, 1999 use a fully nonlinear extended Boussinesq type model FUNWAVE (Wei *et al.*, 1995) to simulate wave induced nearshore circulation. They used laboratory data set from Haller *et al.*, 2002, and found good agreement between simulated and measured parameters. They found that rip currents are unstable, which results in an oscillating of the rip. They observed wave refraction due to the rip currents from the model results. The refraction causes non-uniformity of radiation stress in alongshore direction. In addition, vortices on the bar crest also observed, which are due to non-uniformity of wave breaking, caused by either wave refraction by the rip currents or bathymetric perturbation.

Haas *et al.*, 2003 analyse rip current systems generated by channels over alongshore bar, using quasi 3D numerical model SHORECIRC (Van Dongeren *et al.*, 1994; Putrevu and Svendsen, 1999), again using physical model from Haller *et al.*, 2002 as comparison. In general, they found that simulation results from the model confirm well with laboratory data. Moreover, they strengthen previous findings by some researchers that wave-current interaction is an important feature in rip current systems.

Many publications put emphasis on the kinematics and characteristics of rip currents for open beach environments, while there are fewer for wind sea dominated ones. A study focusing on wind sea dominated beaches can be found in Winter *et al.*, 2014, who performed numerical modelling for rip currents in Egmond, The Netherlands, which is the area of interest of this proposed study. They constructed a model for XBeach and fed it using real bathymetry surveyed using jet ski, and validated the result against GPS drifter measurements. She concluded that in Egmond, there are in general three types of rip currents flow pattern: local one-sided circulation cell, strong offshore drift and then diverted by alongshore current and meandering flow pattern. The first type is related to weak rip current strength, while the second one corresponds to strong rip strength. Rip current magnitudes are in range of 0.18 m/s to 0.6 m/s (the highest during the field campaign, which corresponds to highest wave height recorded). Some key points addressed are: firstly, the model sufficiently reproduces rip flow and conforms relatively well with GPS drifter measurements; secondly, wave current interaction is found to be crucial for the modelling of rip flow; thirdly, cross shore resolution of the model has to be high enough (e.g. 5m) in the nearshore zone to resolve flow around the circulation cell; and fourthly, the tidal current has to be take into account since this is a dominant force in Egmond beach nature as well as important in impacting the flow around the circulation cell.

2.3 Rip current prediction systems

2.3.1 Data driven approach

Several approaches have already been established by some researchers in providing prediction of risk of rip current events for public safety. One method to be mentioned firstly is the LURCS scale method, which stands for LUshine Rip Current Scale (Lushine, 1991). During this work, the number of surf drownings related to rip currents was examined at Dade and Broward County in Southeast Florida. Using the drowning data logs, a correlation between rip currents events, local wind direction and speed, and tidal heights were established. It is found that rip currents are strongly related to onshore wind flow, and the danger is greatest around the low tide. Out of this relation, an experimental scale then developed, based on the relation between rip currents, wind, tide, and swells. In principle, the scale categorizes rip dangerous from zero to five. Category zero simply means no weather related rip current danger, and category five indicates high danger.

On the other hand, Lascody, 1998- following up the work of Lushine, 1991-, reported that instead of being related to onshore wind, rip currents are strongly related to high period of swells impacting the beaches. According to Lascody, 1998, almost 80% of rip rescues are due to swells, and another 20% can be attributed to onshore wind. This was found using data recorded from 1986 to 1995 for three different locations. Based on this work, the so called LURCS checklist was developed (see Figure 2-2 for an example). Engle et al., 2002 make further improvement to the method. Using lifeguards rescue data from Daytona Beach in Florida, it is found that the frequency distribution of rip currents rescues has distinct peaks during shore-normal wave incidence and low tide, deep water wave heights in order of 0.5 to 1 meter and wave period of 8 to 10 seconds (see Figure 2-3).

East central Florida LURCS checklist and computed value (shaded) for 5/31/97.

1. WIND FACTORS	MOST FAVORABLE FOR RIP CURRENTS	MOST FAVORABLE FOR LONGSHORE CURRENTS
SPEED / DIRECTION	(40-110◆)	(120-160◆, 340-30◆)
5 kt	0.5	0.0
5-10	1.0	0.5
10	1.5	1.0
10-15	2.0	1.5
15	3.0	2.0
15-20	4.0	3.0
20	5.0	4.0
20-25+	5.0	4.0
	WIND FACTOR	0.5

2. SWELL FACTORS		
a)	SWELL HEIGHT	SWELL HEIGHT FACTOR
	1 ft	0.5
	2	1.0
	3-4	2.0
	5-7	3.0
	8-10	4.0
b)	SWELL PERIOD	SWELL PERIOD FACTOR
	7-8 sec	0.5
	9-10	1.0
	11-12	2.0
	>12	3.0
c) SWELL HEIGHT FACTOR + SWELL PERIOD FACTOR = SWELL FACTOR		4.0

3. MISCELLANEOUS FACTORS	
If astronomical tides are higher than normal (i.e., near full moon), add 0.5	
If previous day Wind Factor or Swell Factor greater than or equal to 2.0/1.5, respectively, add 0.5	
MISCELLANEOUS FACTOR	0.5

4. TODAY'S RIP CURRENT THREAT is a summation of the 3 factors.	
LONGSHORE / RIP CURRENT THREAT	5.0

5. If RIP CURRENT THREAT is 3.0 - 4.0** (2.5 - 3.5 ** on weekends/major Holidays): issue statement for greater than normal threat of rip currents.

If RIP CURRENT THREAT is 4.5 - >5.0 ** (4.0 - >5.0 ** on weekends/major Holidays): issue statement for much greater than normal threat of rip currents and/or heavy surf. ** (and it looks reasonable, e.g., an arctic outbreak, rainy day, hurricane, etc. is not occurring)

Figure 2-2: An example of LURCS checklist, showing potential rip current events (Lascody, 1998)

2.3.2 Process based model approach

A different approach in providing alerts to swimmers is taken by Alvarez-Ellacuria *et al.*, 2009, called the Hazard Alert System (HAS). This system uses the operational wave forecasting system of the Spanish Harbour Authority for offshore data source, and then transforms the condition towards nearshore. A nearshore database consisting of wave height-period-direction classification is built (see Figure 2-4). The hazard level is determined based on nearshore wave conditions derived from the database. Therefore, wave conditions which are considered hazardous are defined based on the incident wave angle around the normal of the beach. Angles of approach can be different per beach, and therefore actual beach state based on lifeguards input, aerial photographs and sediment size, is taken into account. The hazard level then will be determined based on wave heights and angle of approach. Information from the HAS prediction system is sent to the lifeguards via short message system and will be published on the local authority's website. Alvarez-Ellacuria et al., 2009 indicate, to have the system work adequately, field observations are required to validate the system and inputs from the local lifeguards regarding beach morphology dynamics is essential for the system. They also mention a need of autonomous observations (e.g. remote video) to further improve the system.

Modified ECFL LURCS Checklist

Example computations appear in **bold**.

Wave Period			Wave Direction		
Period, T (s)		Factor	Direction, θ (deg)		Factor
T < 6		0	θ < -35 or θ > 20		0
6 <= T < 9		**0.5**	-35 <= θ < -30 or 20 >= θ > 15		1
9 <= T < 11		1	-30 <= θ < -25 or 15 >= θ > 10		2
11 <= T < 12		2	-25 <= θ < -15 or 10 >= θ > 5		3
T >= 12		3	**-15 <= θ <= 5**		**4**
Wave Period Factor =		**0.5**	**Wave Direction Factor =**		**4**

Wave Height		
Height, Ho (ft)		Factor
Ho < 1		0
1 <= Ho < 2		0.5
2 <= Ho < 3		**1**
3 <= Ho < 5		2
5 <= Ho < 8		3
Ho >= 8		4
Wave Height Factor =		**1**

Tide		
Tide, h (m)		Factor
h > -0.2		0
- 0.5 < h <= -0.2		**1**
- 0.75 < h <= -0.5		2
h <= -0.75		1
Tidal Factor =		**1**

Directional Spreading	
Dspr, (θ°)	Factor
θ > 35	0
30 < θ <= 35	**3**
θ < 30	4
Dspr Factor =	**3**

Sum the factors: The Modified ECFL LURCS rip current threat =	**9.5**

Figure 2-3: Modified LURCS checklist (Engle et al., 2002)

A more engineering approach is taken by Alvarez-Ellacuria et al., 2010. They develop an operational wave and current forecasting system for the north eastern part of Mallorca Island, Spain. They make use of a 2DH Navier-Stokes model for the nearshore to generate surf zone waves and current predictions. This model is forced by an onshore wave propagation model. This model solves the mild slope equations, and derives its two dimensional wave spectra input from deep water wave prediction model WAM Cycle 4. This wave model is forced by wind field from HiRLAM. It is reported that during the year of 2005-2007, the correlation between model prediction and measurements for the wave heights and mean period are 86% and 78% respectively. Moreover, the rms errors for these parameters are 0.37 and 1.32 respectively. To have an insight of the model performance in detecting rip currents, an aerial photograph of the beach is utilized. Rip current locations are visually identified by this seaweed, or a line of foam, or debris moving steadily seaward. However, this operational activity requires expensive and accurate bottom mapping/bathymetry for at least once every 2 months on average. This is due to the dynamics of the sand bar and the rip channels. Therefore Alvarez-Ellacuria *et al.*, 2010 suggest to further investigate utilization of video technique to update the bathymetry.

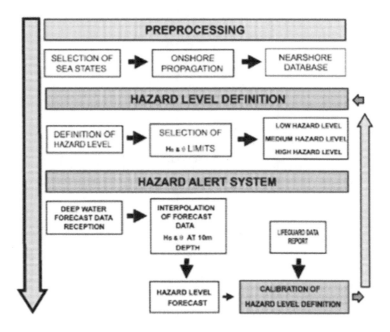

Figure 2-4: Hazard Alert System (HAS) forecasting system (Alvarez-Ellacuria *et al.*, 2009)

A tool similar to the one previously mentioned is developed by Austin *et al.*, 2013 for Perranporth beach, UK. They use an operational regional wave model to force a local model for coupled wave propagation and tidal flow. The system was tested using surveyed bathymetry and validated using measured Eulerian field data. The system was also tested using forecast forcing, to provide an example of forecast-mode application. They emphasize that a key requirement for an operational rip risk prediction tool is representative bathymetry, which needs to be updated frequently. To this end, they suggested to involve nearshore bathymetry estimated using video technique, which is further addressed by Van Dongeren *et al.*, 2013. To tackle the need of frequently updated bathymetry, Kim *et al.*, 2011 delivered a fully process-based rip current prediction tool,

SADEM, comprising a wave propagation model coupled with sediment transport module to update bed level changes. They tested the system for Haeundae Beach, in Korea, and obtained a promising tool to be developed and applied for forecasting of dangerous rips.

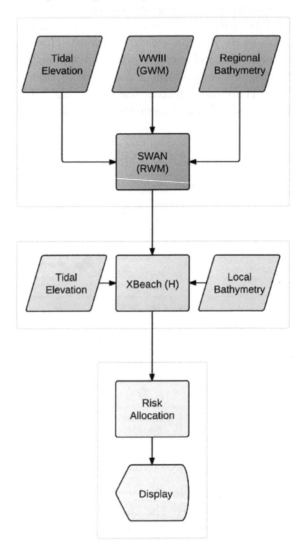

Figure 2-5: The flow diagram of the Rip Risk Prediction Tool (RRPT, Austin *et al.*, 2013)

There are some scientific efforts which are tried to locate rips based on remote sensed technology, for instances video (Ranasinghe *et al.*, 1999; Bogle *et al.*, 2000; Gallop *et al.*, 2009), and high frequency radar (Kohut *et al.*, 2008). Detection of rip positions using video technique is reported to be successful. However, the detection for the rips is not in operational mode, providing forecast information about rip threats. A different method was presented by Kohut *et al.*, 2008, where they operated a high frequency radar system to predict the daily situation of surface currents at Mid Atlantic Bight, which actually is more on a regional spatial scale, rather than focused on a specific coastal area.

2.4 Conclusion

Rip currents have long been a research topic worldwide. While literature is available discussing characteristics of rip currents on open beaches, little is found for wind sea dominated beaches. In this thesis, through numerical modelling, the kinematics of rip currents at Egmond aan Zee beach, located on the Dutch coast, will be investigated. In addition, the implications for swimmer safety will also be addressed.

Systems to predict rip currents occurrence have been developed all around the world. Data driven prediction systems like the one proposed by Lushine, 1991 and Lascody, 1998 require a long record of rip incidents, which for Egmond aan Zee is not available. Moreover, these prediction systems do not take into account the nearshore bathymetry condition as a variable. On the other hand, process-based approaches like the ones from Alvarez-Ellacuria *et al.*, 2010 and Austin *et al.*, 2013 show a promising and efficient way of providing forecast, and suggest an improvement by using up to date bathymetry derived from model-data assimilation technique. In this thesis furthering the work presented by Van Dongeren *et al.*, 2013, a combination of numerical prediction tools and nearshore bathymetry from video, will be demonstrated in order to provide rip current forecasts.

3 Coastal operational model – CoSMoS – system set up and validation[3]

3.1 Introduction

Knowledge of the actual condition of nearshore and coastal hydrodynamics is an essential point in coastal risk management and monitoring activities. Using this knowledge, the risk of coastal hazards, such as coastal inundation, beach and dune erosion, and rip currents can be predicted and mitigated. To this end, a coastal operational model system can serve as a key tool in providing recent and up-to-date information about the hydrodynamic and morphodynamic state of the coast. The output from such a model system can be valuable for coastal stakeholders and decision makers. In this paper validation result of CoSMoS (Coastal Storm Modelling System, Baart *et al.*, 2009; Van Ormondt *et al.*, 2012; Barnard *et al.*, 2014), an operational coupled wave and tide-surge modelling system, is presented. The CoSMoS model system is generic and its application is not limited to storm events, but can also be used for daily condition applications such as operational workable weather forecast for the marine and offshore industry as well as rip current predictions for swimmer safety application. Nowadays, many models are already available with which coastal processes and circulation can be simulated (Van Dongeren *et al.*, 1994; Wei *et al.*, 1995; Roelvink *et al.*, 2009). Model systems like CoSMoS can provide such coastal models with boundary information from larger area models in an efficient way and with low logistical efforts. In addition, meteorological data as input for CoSMoS can be obtained from well-established meteorological models, most of

---

[3] This chapter is based on Sembiring, L., van Ormondt, M., van Dongeren, A. and Roelvink, D., 2015. A validation of an operational wave and surge prediction system for the Dutch coast. *Nat. Hazards Earth Syst. Sci.*, 15(6): 1231-1242.

which are run in operational mode, e.g. GFS (Global Forecast System, operated by National Oceanic and Atmospheric Administration, USA), HIRLAM (High resolution limited area model, Unden et al., 2002), and ECMWF (European centre for medium-range weather forecast, Janssen et al., 1997).

Application of such model systems for wave forecasting has been demonstrated previously. On a global scale, Hanson et al., 2009 presented a skill assessment of three different regional wave models: WAM (Gunther, 2002), WAVEWATCH III (Tolman, 2009), and WAVAD (Resio and Perrie, 1989). They performed multi-decadal hindcast for the Pacific Ocean, and found that in general, all three models show good skill, with WAVEWATCH III performing slightly better than the other two. For semi-enclosed basin scale modelling, some studies show that wave conditions can be well simulated in serious storm events (Cherneva et al., 2008; Bertotti and Cavaleri, 2009; Bertotti et al., 2011; Mazarakis et al., 2012). However, the quality of the wave model decreases substantially when the wind condition shows strong temporal and spatial gradients. This is particularly true for enclosed basins where an underestimation of wind speed by the atmospheric model is often found (Cavaleri and Bertotti, 2004; Cavaleri and Bertotti, 2006; Ponce de León and Guedes Soares, 2008). In addition, wave models are very sensitive to small variations within the wind fields which act as forcing input (Bertotti and Cavaleri, 2009). Moreover, Cherneva et al., 2008 reported an underestimation by the WAM Cycle 4 model of significant wave heights in the case of low wind energy input and during combined swell-wind wave conditions. In contrast, the model shows relatively better performance for the case of high energy input. For the North Sea area, Behrens and Günther, 2009 demonstrated that WAM model that covers the North Sea and Baltic Sea, was capable to provide forecast 2 days ahead of winter storm Britta and Kyrill in 2006 with satisfactory results. As improvements to the model, they suggest further development on the atmospheric model (developed by German Weather Service). In addition to that, the model does not take into account depth induced wave breaking as one of the source terms in the model equation, which is a required further improvement necessary for a nearshore prediction system.

The application of model systems for storm surge and tide prediction has also been demonstrated. For the Netherlands, Verlaan et al., 2005 make use of the hydrodynamic model, DCSM (Dutch Continental Shelf Model, Gerritsen et al., 1995) and forced it using meteorological model HIRLAM, to provide tide and storm surge forecasts. Moreover, they implemented an update to the Kalman filter configuration to improve quality of forecast, initially implemented by Heemink and Kloosterhuis, 1990. This approach now has been applied in the operational prediction system Delft-FEWS (Delft- Flood Early Warning System, Werner et al., 2013), which has flexibility in integrating different models and data in a comprehensive way to provide forecast information. Specifically for coastal forecasting, De Kleermaeker et al., 2012 present an operational model system for the Dutch Coast under the framework of the FEWS system, combining data from different sources to provide a reliable forecast. They use the hydrodynamic model of Delft3D-FLOW to compute tides and surge and SWAN (Booij *et al.*, 1999) for the waves. Preliminary results from the model system give a wave height bias of 10 % and root mean square error of 3.7 cm for the water levels.

Coupling wave-tide-surge modelling has been found to be an important key to improve prediction skill of water level and waves (Wolf, 2008; Brown and Wolf, 2009). Application of this coupling approach for Mediterranean Sea using a 3-D finite element hydrodynamic model SHYFEM shows that the modelling system Kassandra predicts water level and wave height very well with the root mean square error range of ~ 4 until 8 cm and ~22 until 33 cm for total water level and wave height respectively (Ferrarin *et al.*, 2013). Similar approach has also been implemented for The Irish Sea by Brown *et al.*, 2010, where they use one way nested approach of North Atlantic model and the Irish Sea region. Using WAM as spectral wave model and the POLCOMS model (Proudman Oceanographic Laboratory Coastal-Ocean Modelling System) as tide-surge model, they show that the coupled scheme performs very well at predicting total water level and waves. They make use of statistical measure called Percentage Model Bias (PBias) to evaluate the skill of the model, and found that model system gives PBias of 14% until 37% for significant wave height (Brown *et al.*, 2010 classify the PBias as: less than 10% excellent, 10-20% very good, 20-40% good, and greater than 40% poor).

Here, the CoSMoS application to the North Sea basin and validation will be presented. It is a coupled wave-tide-surge modelling approach with nested spatial domain starting from the global model to the North Sea. The validation will be on the wave parameters, water level, and surge elevation simulated by the model, focusing on the North Sea.

21

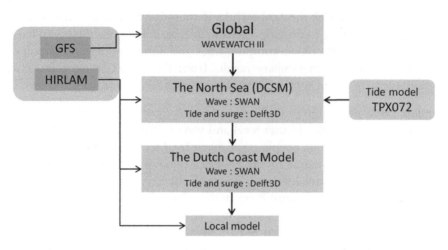

Figure 3-1: Model and domain structure within CoSMoS

Table 3-1: Model set up for DCSM

	SWAN	Delft3D Flow
Grid size	~ 15 x 20 km²	~ 7.5 x 10 km²
Open boundary	WAVEWATCH III Global	Tide model TPX072
Meteo- input	HIRLAM	HIRLAM
Source term:		-
Wind growth	van der Westhuysen *et al.*, 2007	
White capping	van der Westhuysen *et al.*, 2007	
Bottom friction	JONSWAP (Hasselman *et al.*, 1973)	
Depth induced breaking	Battjes and Janssen, 1978	

3.2 The CoSMoS model system

The CoSMoS system is set up for the North Sea basin where regional wave and tide-surge models are integrated with local (high resolution) models. The model and domain structure of CoSMoS appears in Figure 3-1. For the global model, Wave Watch III (WW3 from here on) is used and forced by six hourly GFS meteo, with resolution of 1x1.25 degree. This global model generates 2D wave spectra as output and will be used as swell boundary conditions for the nested models, as indicated in Figure 3-1 by the arrows.

The next nested model is the Dutch Continental Shelf Model (DCSM) which comprises the wave model SWAN and the tide-surge model Delft3D-FLOW.

The spatial resolution of the surge model is approximately 7.5 x 10 km, while the resolution of the wave model is approximately 15 x 20 km. The surge model is driven by meteo data from HIRLAM, which are 3 hourly data with approximately 9 x 14 km resolution. In addition, amplitude and phase of several relevant tidal constituents are assigned using the tide model TPX072 (Egbert and Erofeeva, 2002). The SWAN model is forced by wind field from HIRLAM and the swell boundary conditions (sections indicated by red lines in Figure 3-2) are obtained from the global WW3 model. The models are run simultaneously, allowing for wave-tide-surge interactions. The model set up of the DCSM model within CoSMoS is summarized in Table 3-1, and model version and the parameter settings used in the model are presented in Table 3-2. For the wave model, white-capping is modelled based on van der Westhuysen *et al.*, 2007, bottom friction formula is from Hasselman *et al.*, 1973, and depth induced breaking model is from Battjes and Janssen, 1978. For the Delft3D FLOW model, uniform bed roughness coefficient is used, and wind drag coefficient is determined by a linear function of three break points and the corresponding wind speed (Delft3D FLOW User Manual). In this paper, The Dutch Coast Model and the Local Model will not be part of the analysis as we will focus on the application of CoSMoS for the North Sea.

Table 3-2: Model version and parameter settings

SWAN, model version 4072	
Source term	Value
White-capping (van der Westhuysen *et al.*, 2007)	
B_R (threshold saturation level)	0.00175
C'_{ds} (proportionality coefficient)	0.00005
Bottom friction (Hasselman *et al.*, 1973)	
C (bottom friction coefficient)	0.067 m^{2}s^{-3}
Depth induced breaking (Battjes and Janssen, 1978)	
Gamma (breaker parameter)	0.73
Alpha (dissipation coefficient)	1.0
DELFT 3D FLOW, model version 4.01	
Bed roughness Chezy coefficient	90 m$^{1/2}$s^{-1}
Wind drag coefficient A, B, C	0.00063, 0.00723,
(see Delft3D User Manual)	0.00723

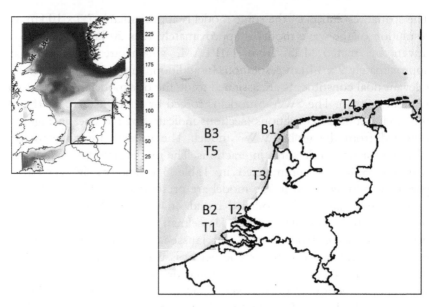

Figure 3-2: The North Sea (left), with the colour shading represents depth. Red lines indicate the swell boundary for the wave model. The Dutch Coast (right), and stations used in the analysis (see Table 3-3)

The system is designed in a MATLAB platform, where the initiation and operational run are performed every 12 hours, and managed by so called timer loop. In Figure 3-3, the workflow of CoSMoS is presented. There are two timer loops in the system that dictates the operational run. First, the main loop, which defines the starting time and end time of the model run, triggers the overall initiation of the system, and downloads necessary wind and air pressure data to be used by the models. The second time loop is the model loop, in which model runs will be executed in sequence, starting from global and regional models followed by higher-resolution models. Downloaded forcing data and simulation results from the models are stored on a local OPeNDAP server (OPeNDAP: Open-source Project for a Network Data Access Protocol, Cornillon *et al.*, 2003).

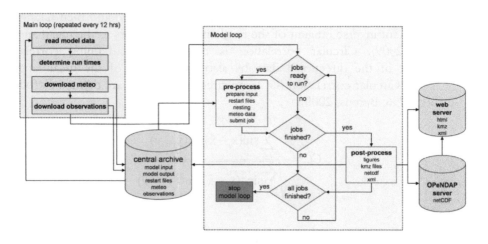

Figure 3-3: The workflow of CoSMoS (Van Ormondt *et al.*, 2012)

3.3 Model system validation

3.3.1 Data and method

The validation is carried out by comparing simulated parameters obtained from the model with the observed ones as ground truth. Statistical error measures are used to quantify the error between simulated and observed data. Here, root mean square error (e_{rms}), bias, and normalized error (e_{norm}) are used. The expressions are as follows:

$$e_{rms} = \left[\frac{1}{N} \sum_{i=1}^{N} (x_i - y_i)^2 \right]^{\frac{1}{2}} \qquad \text{3-1}$$

$$bias = \frac{1}{N} \sum_{i=1}^{N} x_i - \frac{1}{N} \sum_{i=1}^{N} y_i \qquad \text{3-2}$$

$$e_{norm} = \frac{\left[\frac{1}{N} \sum_{i=1}^{N} \left[(x_i - \bar{x}) - (y_i - \bar{y}) \right]^2 \right]^{\frac{1}{2}}}{\bar{y}} \qquad \text{3-3}$$

In the equation 3-1 until 3-3, N is the length of the time series parameter, x and y are simulated and observed parameter respectively, $\bar{x}$ and $\bar{y}$ are mean value of x and y respectively. For directional data, circular correlation and circular bias are used as statistical error measures. The circular bias is defined by subtracting the mean angular of the computed parameter from the mean angular of the observed. Here, the mean angular is computed by transforming the directional data into two vector components with magnitude of unity, and then taking the four quadrant inverse tangent of the resultant of the vectors as the mean angular (Berens, 2009). Circular correlation is computed by defining correlation coefficient of the directional data by also making use of the mean angular measures. Circular correlation (CC) and circular bias (CB) are defined as follows (Fisher, 1996; Berens, 2009):

$$CC = \frac{\sum_{i=1}^{N} \sin(x_i - \hat{x})\sin(y_i - \hat{y})}{\sqrt{\sum_{i=1}^{N} \sin^2(x_i - \hat{x})\sin^2(y_i - \hat{y})}} \qquad 3\text{-}4$$

$$CB = \hat{x} - \hat{y} \qquad 3\text{-}5$$

$$\hat{x} = \arctan(R) \qquad 3\text{-}6$$

$$R = \frac{1}{N}\sum_{i=1}^{N}\begin{bmatrix} \cos(m_i) \\ \sin(n_i) \end{bmatrix} \qquad 3\text{-}7$$

In the equation 3-4 until 3-7, $\hat{x}$ and $\hat{y}$ are the mean angular of simulated and observed parameter respectively, (m,n) is the plane component of the directional data (unit vector), and R is the mean resultant vector.

Table 3-3: Stations used in validation

Station	Name	Abbreviation	Type
B1	Eierlandse Gat	EIELSGT	Directional wave buoy
B2	Euro platform	EURPFM	Directional wave buoy
B3	K13 platform	K13APFM	Directional wave buoy
T1	Euro platform	EUR	Tidal gauge
T2	Hoek van Holland	HvH	Tidal gauge
T3	IJmuiden	IJM	Tidal gauge
T4	Huibergat	HUI	Tidal gauge
T5	K13 platform	K13	Tidal gauge

As the ground truth, data from deep water directional wave buoys and tidal gauge record are used, of stations located near the Dutch Coast. Three wave buoys are considered: Eierlandse Gat (EIELSGT), K13 Platform (K13APFM), and Euro Platform (EURPFM). For water levels, five gauges are used: Euro platform, Hoek van Holland, IJmuiden, Huibergat, and K13 platform (see and Figure 3-2). Data obtained from the buoys are processed, stored and retrievable as wave energy density, mean wave direction, and directional spreading as function of frequency, rather than the full 2D spectra. Therefore, quasi 2D wave energy spectrum is constructed using following expressions:

$$E(f,\theta) = E(f) \cdot D(\theta, f) \qquad \text{3-8}$$

Where $E(f, \theta)$ is the wave energy as a function of frequency and direction and D is directional spreading. Since:

$$\iint E(f,\theta)dfd\theta = \int E(f)df \qquad \text{3-9}$$

, therefore

$$\int D(\theta, f)d\theta = 1 \qquad \text{3-10}$$

For the directional spreading, a normal distribution function is used:

$$D(\theta, f) = \frac{1}{\sigma\sqrt{2\pi}} \exp\left[\frac{[\theta - \theta_0(f)]^2}{2\sigma(f)^2}\right] \qquad \text{3-11}$$

where $\theta_0(f)$ is mean direction as function of frequency, σ is directional spreading as function of frequency, and θ is the running wave direction. Here we have to keep in mind that the shape of the directional spreading function is assumed to be Gaussian and directional bimodality is not significant over the period of the data (Longuet-Higgins and Stewart, 1963; Wenneker and Smale, 2013).

Since the wave model SWAN returns output as two-dimensional wave spectra as well, consistent parameter definitions can be used for both simulated and observed data. The integral wave parameters then will be calculated as follow:

$$H_{m0} = 4\sqrt{\iint E(f,\theta)dfd\theta} \qquad \text{3-12}$$

$$T_p = \frac{1}{f_p} \qquad \text{3-13}$$

For the mean wave direction, formula from Kuik *et al.*, 1988 is used:

$$\theta_{mean} = \tan^{-1}\left[\frac{\iint \sin\theta E(f,\theta)dfd\theta}{\iint \cos\theta E(f,\theta)dfd\theta}\right] \qquad 3\text{-}14$$

For analysis purposes, in addition to bulk wave parameters, wind sea and swell components will also be computed. To this end, an algorithm will be applied on the total wave spectrum to differentiate between energy that belongs to wind sea and swell. The algorithm will largely follow Hanson *et al.*, 2009 and Portilla *et al.*, 2009. Here, for simplification it is assumed that the total energy content in the spectrum only consists of one system of wind sea and one system of swell. The demarcation line between sea and swell is defined as:

$$f_c = \frac{g}{2\pi}\left[\alpha U \cos(\delta)\right]^{-1}$$

$$0 \le \delta \le \frac{\pi}{2} \qquad 3\text{-}15$$

where f_c is the critical frequency, α is a constant, U is wind speed, and δ is the angle between wind sea and the wind. Energy content above this line will be counted as wind sea spectrum while below it is swell. The calibration parameter α of 1.8 is used.

3.3.2 Results and discussion

Hindcast

A hindcast was performed for the calendar year of 2009, where the forcing of the model system is provided by the analysed wind fields. In general, the year 2009 exhibited typical yearly wave conditions for the North Sea without any particular significant storm event. As we are also interested in operational daily performance rather than specific extreme event analysis, this particular year is thus a representative one. Weekly averaged observed significant wave height varied from 0.5 m during week 27 (the month of July) up to 2.8 m during week 48 (the month of November). For the latter period, the maximum observed wave height was 4.79 m. These wave heights are approximately in the same range for the three buoys considered.

Water level and surge validation

Model performance for water level and the surge is analysed by comparing both the simulated tidal signal with the observations. The surge levels were determined by subtracting astronomical prediction from the tidal signal. Results show that simulated tide and surge levels are in good agreement with observations. Figure 3-4 presents water levels plot for location IJmuiden during the storm that occurred in the last week of November 2009, where the computed water level (green) and observed water level (blue) elevates from astronomical prediction (grey). This water level raising is clearly seen from the observed surge level (black) which in a good agreement with the computed surge (red). Monthly error plots for the surge levels appear in Figure 3-5. Root mean square error (left panel) vary from 0.09 m for station K13 platform for the month of September, until 0.21 m for station Europlatform for the month of March. For the bias (right panel), the highest positive value is found at station Huibertgat for the month of February with a bias of 0.12 m, while the strongest negative bias of -0.08 m is given by the station K13 platform for the month of June. The relatively higher surge rms error coincides with the winter period where stormier and higher wind speeds are expected. This seasonal trend is clearly seen from Figure 3-5, where all the tidal gauges considered show a similar tendency of lower rms error during summer months with relatively higher rms error during winter months. An exception is station K13 platform, where the rms error is relatively constant around 0.05 to 0.1 m over the year. This is due to the location of K13 platform which is relatively far offshore, which makes it less prone to the variability of wind driven surge. In addition, the results also show that stronger positive bias is found mainly during winter period, while during the calmer months the absolute bias is relatively smaller with a tendency of being negative. Overall, the surge is in good agreement with observation over the year.

A tidal analysis was performed towards the computed water level where the amplitude and the phase of several relevant tidal constituents are compared with observations. Figure 3-6 presents a bar-plot of the tidal amplitude for six most dominant tidal constituents at the Dutch Coast. For the most important constituent, the M2, model (red) tends to slightly over predict the observations (black), except for stations K13 platform. Relative error is from 0.4% for IJmuiden to 9% for K13 platform. Similar to Figure 3-6, in Figure 3-7 the bar-plot of tidal phase is presented. Tidal phase is predicted well by the model, with a higher error tendency appears in diurnal constituents K1 and O1. The absolute differences between computed phase and observations for the most important constituent M2 is 2 degrees for K13 platform until 8 degrees for IJmuiden.

The error measures shown by the model verify that tide and total water level can be predicted very well by CoSMoS model. For comparison purpose with similar coupled wave and tide-surge modelling approach, error metrics Cost Function CF and Pbias (after Brown et al., 2010) of the CoSMoS model in predicting total water level and the surge were calculated. Over the whole tidal gauges used in the analysis, CoSMoS gives a PBias range of -1.68% until -13.75% for total water level and -5.9% until -41% for the surge. Eleven year hindcast of POLCOMS-

WAM model applied for the Irish Sea by Brown *et al.*, 2010 give ~ -1% until -14.9% for total water level and -3% to -52% for the surge. Note that in Brown *et al.*, 2010, negative value of PBias suggesting an overestimation tendency and vice versa, in contrast with one used in this paper. Error metric Cost Function CF also gives good performance evaluation for CoSMoS. For total water level and surge, the ranges of CF value are 0.18 to 0.27 and 0.61 to 0.74 respectively (CF < 1 means the model has a chance of predicting skill, CF <0.4 means the models simulated the variables well).

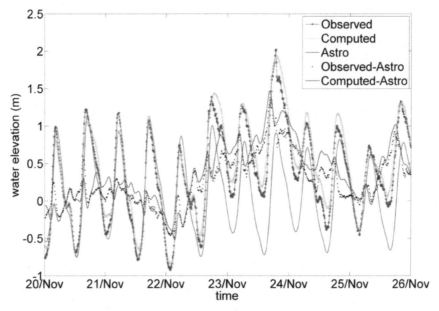

Figure 3-4: Water elevation for IJmuiden during the storm in November 2009, blue: observed water level, green: computed water level, black: observed surge, red: computed surge, grey: astronomical prediction.

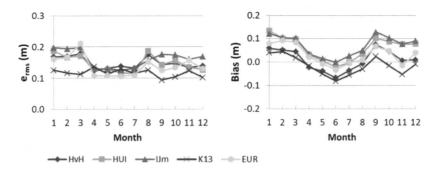

Figure 3-5: Monthly root mean square error (left) and bias (right) of surge elevation.

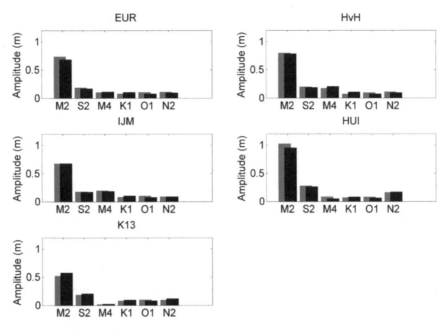

Figure 3-6: Tidal amplitude comparison for six tidal constituents, black: observation, red: computed

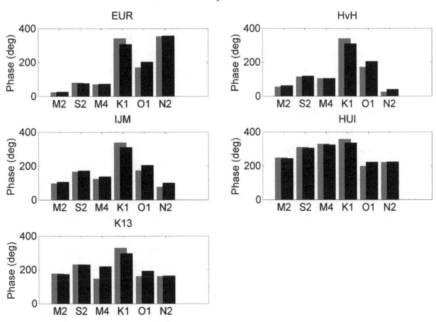

Figure 3-7: Tidal phase comparison for six tidal constituents, black: observation, red: computed

Wave validation

In general, the wave model shows good skill in reproducing the wave parameters for the hindcast period. Examples of time series plots with simulated and observed wave parameters for K13 platform are presented in

Figure 3-8, where a storm occurred at the end of November. The wave height, peak period and mean wave direction computed by the wave model (blue) are in good agreement with the observations (red dots) over the two-week period that is shown, with a slight overestimation tendency of the observed wave height by the model. Monthly wave parameter errors are given in Figure 3-9 for all the buoys considered. Over the whole year, the root mean square error (Figure 3-9a, left panel) ranges from 0.16 m (Eierlandse Gat, month of May) until 0.35 m (Europlatform, month of November). The normalized error for the wave height (Figure 3-9a, right panel) is consistently below 0.3 with bias in a range of -0.15 m to 0.15 m (Figure 3-9a, middle panel). For the peak wave period (Figure 3-9b), the rms error ranges from 1 second at the Europlatform buoy for the month of November, to 2.1 seconds at Eierlandse Gat for the month of September (Figure 3-9b, left panel). The bias for the peak wave period is consistently negative (Figure 3-9b, middle panel), ranging from -1.05 second up to -0.08 second for all buoys considered, suggesting that the model gives a consistent underestimation of peak wave period over the hindcast time. For the mean wave direction, model results show a significant agreement with the observations (Figure 3-9c). The circular correlation varies from 0.7 at Europlatform and Eierlandse Gat during the month of January, up to 0.9 for the most of the months, showing a high correlation between the model result and the observation. Bias varies between −16 degree and 9 degree with most of the months give bias approximately +/- 10 degrees.

The error measures show consistently low values, which are within the range of error metrics reported by Ferrarin et al., 2013 which uses a similar approach of coupling wave-tide-surge modelling. In addition, these results also show comparable skill with ones from Brown et al., 2010 whose applying similar coupling method to the Irish Sea. The error metrics Cost Function CF for CoSMoS was calculated, and a relatively constant value of ~0.3 was obtained for Hs, and ~0.7 for Tp, over the 3 buoys analysed, whilst Brown et al., 2010 obtained the CF values in a range of 0.4 to 0.6 and 1.4 to 4.0 for Hs and Tp respectively. This suggests that CoSMoS demonstrates very good predictability skill. In Figure 3-10, scatter plots of observed and computed wave height are presented, with different colour indicating different peak wave period, for station K13 platform during relatively calm period (the month of June, Figure 3-10a) and stormy period (the month of November, Figure 3-10b). Six panels scatter plot in (a) and (b) indicate 60 degrees bin of observed incoming wave direction. The results show that the wave model tends to underestimate northerly waves with peak wave period greater than 7 seconds (blue and red scatter dots in Figure 3-10a top-left panel, in Figure 3-10b top-left and bottom-right panel). For semi enclosed seas like the North Sea, the wave climate will be dominated by wind waves while occasional swells can be expected to be present that mainly come from the North, where the shelf is open to the northern part of the Atlantic Ocean. From the results, it is shown that for waves with peak period greater than 7 seconds and coming from the north (between 330 and 30 nautical degree), there is a consistent tendency of the wave model to underestimate the wave height. In contrast, waves that come from west/southwest do not suffer from this underestimation (direction between 150-210 degrees and 210-270 degrees). This tendency is observed in all the buoys considered.

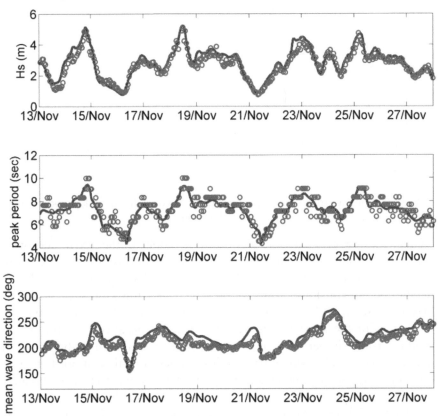

Figure 3-8: Time series of wave parameters, blue: model results, red dots: observations. Wave height (upper panel), peak period (middle), and mean wave direction (lower). Location: K13 platform

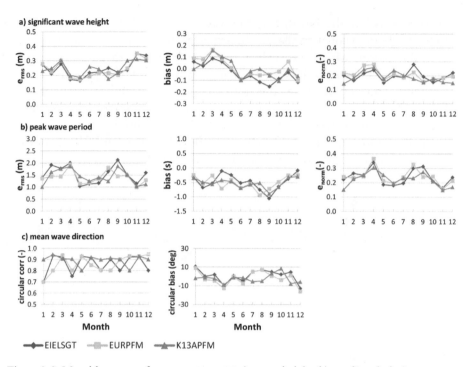

Figure 3-9: Monthly errors of wave parameters, a): wave height, b): peak period, c): mean wave direction

When the wave separation algorithm is applied, the swell and wind sea component results show fairly good agreement with observations. In Figure 3-11 monthly error and bias of swell height component (upper row) and wind sea height component (lower row) are presented. For the wind sea component, the bias is between -0.14 m and 0.19 m (middle panel lower row), which is in contrast with the swell height bias that remains negative over the whole year (middle panel upper row in). This suggests again that the wave model underestimate swells, which may result in underestimation of northerly waves at the Dutch Coast. Lower frequency part of the total wave energy can be an important component not only during storm periods but also for daily normal conditions. For normal conditions, several studies use swell component height as one of the parameters in their statistical prediction system of rip currents, which were built based on the correlation between wave conditions and the number of beach rescue due to drowning (Lushine, 1991; Lascody, 1998; Engle, 2002). This opens doors for the future works on improving the wave model as well as the swell boundary.

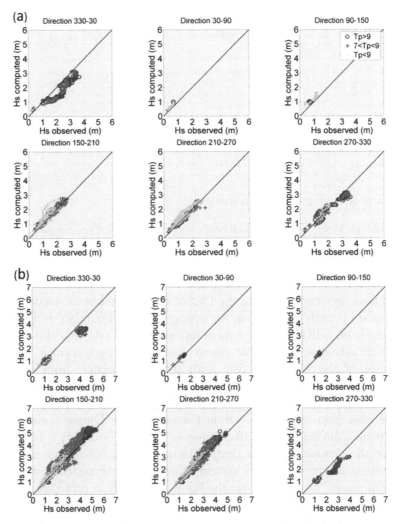

Figure 3-10: Scatter plot of observed significant wave height Hs versus computed, for every 60 degree of observed incoming mean wave direction, for different peak wave period, Tp. Green: observed Hs with Tp<7 seconds, red: 7<Tp<9 seconds, blue: Tp>9 seconds. (a) month of July, (b) month of November. Location: K13 platform.

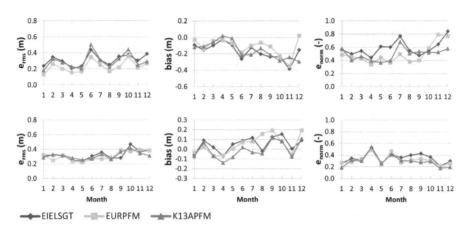

Figure 3-11: Monthly errors of swell height (first row) and wind sea component (second row)

Effect of swell boundary as model system component

The CoSMoS system applies the global WAVEWATCH III model to derive the swell boundary conditions for the DCSM model (red lines in Figure 3-2). Here, a simulation is performed where we ignore the swell boundary and use only wind forcing from HIRLAM as the main input into the wave model to see the effect of such scenario. The result shows that model performance degrades after removal of the swell boundary with different reaction from the model for different periods. Figure 3-12 present time series of wave height with and without the swell boundary for the month of June (left panel) and the month of November (right panel) for station Eierlandse Gat. During the month of June (left panel), the degradation after ignoring the swell boundary is fairly significant, where the simulation result (red) suffers more underestimation compared to default setting of the model system (blue). The normalized root mean square error increases (on average over the buoys) by 3%. In contrast, during the month of November (right panel), the effect of negating the swell boundary is twice less than the month of June, with a 15 % increase in normalized rms error.

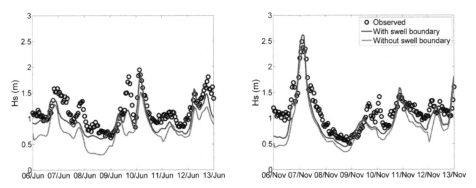

Figure 3-12: Significant wave height Hs for Eierlandse Gat for the month of June (left) and November (right), circle: observed; blue: with swell boundary; red: without swell boundary.

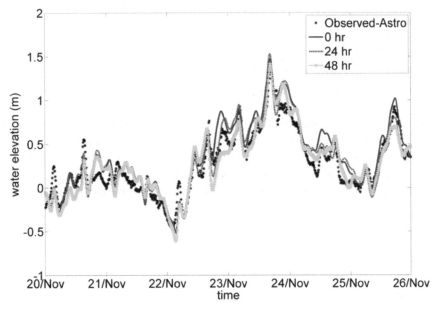

Figure 3-13: Surge elevation for IJmuiden during the November 2009 storm using, blue: analysed HIRLAM; red: 24 hours HIRLAM; green: 48 hours HIRLAM; black dots: observation

Validation in forecast mode

In order to assess the capability of CoSMoS to predict events a number of days into the future, a forecast mode validation is performed where the model system is forced by HIRLAM forecast winds of the year 2009. Two types of forecast HIRLAM wind fields are considered: a 24 hours forecast and a 48 hour forecast. Figure 3-13 shows an example of surge elevation plots for different HIRLAM inputs together with the observations for IJmuiden during November 2009.

Results using the HIRLAM forecasts both for 24 hours and 48 hours ahead (red and green line, respectively) retain good agreement with observations (black dots). The differences are relatively small between results using analysed wind (blue) and results using either the 24 or 48 hour forecast wind. Monthly error plots of surge elevation for different HIRLAM are shown in Figure 3-14 for station IJmuiden. A strong seasonal feature is retained where a relatively lower rms error and bias exhibited during summer season while stronger bias is seen during summer period. This greatest rms error increase is 33% (from analysed to 48 hours) for the month of December. On average, the rms error increases by 11% (from analysed to 24 hours) and 25% (from analysed to 48 hours) for all stations.

For the wave model, time series of wave heights for different HIRLAM forcing is presented in Figure 3-15, for five days in the last week of November for location K13 platform, when the highest weekly-averaged wave heights were observed. The forecast results (red: 24 hours, green: 48 hours) are also in good agreement with observations (black circle). However, there are clear differences, especially with the 48 hour HIRLAM forecasts, while results using analysed and 24 hours HIRLAM look relatively similar. Monthly errors of significant wave height for different HIRLAM are shown in Figure 3-16 for location K13 platform. The rms error tends to increase as the forecast horizon increases. For instance, at K13 platform for the month of November, the rms error was 0.31 m with analysed wind, and further increased to 0.46 m and 0.56 m with 24 and 48 hours winds, respectively. On average, stronger error increments are more prominent during the winter period. In contrast, the monthly-averaged bias does not increase strongly as the forecast horizon increases. On average, the rms error increases by 20% and 40% for 24 and 48 hour forecast, respectively for all locations.

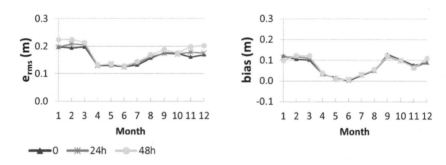

Figure 3-14: Monthly errors of surge elevation for different HIRLAM for location IJmuiden

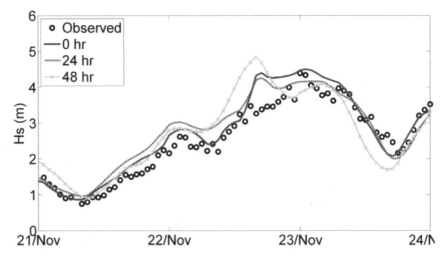

Figure 3-15: Significant wave height Hs for different HIRLAM input, blue: analysed; red: 24 hours forecast; green: 48 hours forecast; black circle: buoy. Location: K13 platform

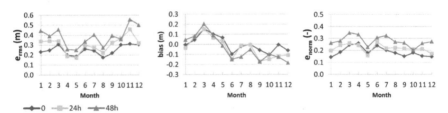

Figure 3-16: Monthly error of significant wave height Hs for different HIRLAM input, location K13 platform

3.4 Conclusions

A CoSMoS model system covering the North Sea is built which employ coupled wave and tide-surge modelling approach to hindcast and forecast wave conditions and water levels in the area along the Dutch Coast. The system is designed in a generic way to accommodate and integrate different regional and local models. A validation is performed of the Dutch Continental Shelf Model (DCSM) covering the North Sea, using wave buoys and tidal gauges available along the Dutch Coast. Hindcast results show that the surge elevation produced by the model is in good agreement with observations with an rms error ranging from 0.09 m to 0.21 m. On average, the model tends to slightly overestimate surge levels, especially during the winter months. For the wave model, simulated wave parameters agree well with observations with relative error of 14% until 30%. However, model tends to underestimate swell height. Using the default settings of the model system (swell boundary is included); a consistent underestimation is found for northerly waves with relatively low frequency, which is also supported by the wave separation algorithm. This suggests room for improvement for the swell boundary conditions on the North of the model domain.

In order to test the CoSMoS system in forecast mode, it has been forced by HIRLAM 24 hours and 48 hours forecasts. The model system is capable of predicting high wave events and storm surge up to two days in advance. However, the performance does degrade as the forecast horizon increases. A smaller error increase is found for the surge elevation than for the wave heights. For surge elevation, on average, the rms error increases by 11% (from analysed to 24 hours) and 25% (from analysed to 48 hours) for all stations. On the other hand, for the wave model, the rms error increases on average by 20% and 40% for 24 and 48 hour forecast, respectively.

To summarize, CoSMoS Dutch Continental Shelf Model is a fit-for-purpose regional model to simulate waves, tide and surge, including the interactions, and to provide boundary conditions for coastal models with which we can use as coastal monitoring tools.

In the future, merging CoSMoS with more integrated operational forecast system like FEWS is recommended in order to have a more accurate prediction of the storm surge level since introducing data assimilation into the system will increase forecast quality.

4 Dynamic modelling of rip currents for swimmer safety on a wind-sea meso-tidal beach[4]

4.1 Introduction

Rip currents are nearshore water flows directed offshore from the surf zone with flow magnitudes up to more than 2 m/s (MacMahan *et al.*, 2006). Because rip currents are the most dangerous coastal hazards for the bathing public, efforts to minimize the risk of drowning have become of great interest. A system that can provide information about where and when rip currents may occur can therefore help provide critical information to lifeguards and the beach-going public. One approach is to numerically predict the occurrence of rip currents. Process-based numerical models have been widely developed and used worldwide in the surf zone (*e.g.*, Noda, 1974; Wu and Liu, 1985; Van Dongeren *et al.*, 1994; Wei *et al.*, 1995; Roelvink *et al.*, 2009). Computational capability has increased very rapidly, which allows us to perform a great number of computations within a reasonable time. With the input from a well-established hydro-meteorological regional model and updated nearshore bathymetry from a remote sensing source, rip current predictions can be generated for a local site in an efficient way (Alvarez-Ellacuria *et al.*, 2010; Kim *et al.*, 2011; Van Ormondt *et al.*, 2012; Austin *et al.*, 2013; Van Dongeren *et al.*, 2013; Sembiring *et al.*, 2014).

[4] This Chapter is based on Sembiring, L., van Dongeren, A., Winter, G. and Roelvink, D. (*In Press*), *Journal of Coastal Research*

In The Netherlands, which has a wind sea dominated coast, the number of incidents due to rip currents is not as large as other typical swell-dominated beaches, (*e.g.,* in the United States or Australia). However, lifeguards on site have shown great interest because risk is considered to increase, with increasing numbers of beach-goers, including people unaccustomed to open coasts (Egmond and s-Gravesande lifeguards, personal communication). Winter *et al.,* 2014 performed the field experiment SEAREX (Swimmer safety in Egmond aan Zee – A Rip current EXperiment) to examine rip current characteristics at Egmond aan Zee, The Netherlands. The results showed that during the field campaign, there were in general three types of flow patterns: local one-sided circulation cells, strong offshore flows that were then diverted by alongshore currents and meandering alongshore flow patterns. They reported that during their experiment, rip current magnitudes were in the range of 0.18 to 0.6 m/s. Rip current occurrence was strongly controlled by wave breaking, and the flow orientation of the rip current outside the surf zone was dictated by the balance between the offshore rip flow and the alongshore tidal current. Based on results from a numerical model with wave group forcing and a stationary tidal current, Winter *et al.,* 2014 showed that obliquely incident waves might not hinder the rip current flow at Egmond because the rip channel is relatively wide (~100 m) compared with the momentum of the wave-driven alongshore current.

Here the focus will be on the bathymetrically controlled rip currents that were observed during the SEAREX field campaign. The objective of this study is to apply a process-based numerical model to predict the initiation, lifespan, and risk indicator of bathymetrically controlled rip currents for Egmond aan Zee. The prediction window for swimming safety purposes can be provided 2 days in advanced following the lead time of the meteorological forecast (*e.g.,* Van Ormondt *et al.,* 2012 for the Dutch Coast application). Moreover, the numerical model used to provide the prediction should be able to simulate conditions including tidal current dynamics. Here, the work of Winter *et al.,* 2014 will be continued, and the numerical model will be run for the continuous 5 days of the field campaign, taking into account the dynamics of the tidal currents, as well as the interaction with the waves. This continuous approach is suitable for operational prediction of rip currents.

The model will be validated against Lagrangian mean flow observations that were measured with GPS-tracked drifters during SEAREX. As objects in water move with the mean flow, the drifters can be considered a proxy for swimmers subjected to (rip) currents, with some limitations as a result of different inertia and submersion characteristics. Inactive human drifters also used during the field work showed floating behaviour similar to the drifters. The bathymetry and the rip channel features are assumed to be invariable over the simulation period.

4.2 Methods

In the following sections, the modelling approach, the field site and the data collected and used during the field work will be described.

4.2.1 Model

An XBeach model was created (Roelvink *et al.*, 2009) to simulate the waves and the nearshore currents. The computation solves simultaneous two-dimensional (2D) horizontal equations of wave action mass and momentum in which interaction between waves and currents can take place. XBeach was originally developed to simulate waves and morphological change during storm conditions. Here, the focus will be on the modelling of the hydrodynamic processes during normal conditions. Sediment transport and bed-level changes are not taken into account in this application because morphological change was small during the experiment.

The depth-averaged, short wave-averaged equations of mass and momentum are used to model the water flow using depth-averaged shallow water equations. Moreover, forcing from waves and wind is also included in the equation. The equations read:

$$\frac{\partial u}{\partial t} + u\frac{\partial u}{\partial x} + v\frac{\partial u}{\partial y} - v_h\left(\frac{\partial^2 u}{\partial x^2} + \frac{\partial^2 u}{\partial y^2}\right) = \frac{\tau_{wx}}{\rho h} - \frac{\tau_{bx}}{\rho h} - g\frac{\partial \eta}{\partial x} + \frac{F_x}{\rho h} \qquad 4\text{-}1$$

$$\frac{\partial v}{\partial t} + u\frac{\partial v}{\partial x} + v\frac{\partial v}{\partial y} - v_h\left(\frac{\partial^2 v}{\partial x^2} + \frac{\partial^2 v}{\partial y^2}\right) = \frac{\tau_{wy}}{\rho h} - \frac{\tau_{by}}{\rho h} - g\frac{\partial \eta}{\partial y} + \frac{F_y}{\rho h} \qquad 4\text{-}2$$

$$\frac{\partial \eta}{\partial t} + \frac{\partial hu}{\partial x} + \frac{\partial hv}{\partial y} = 0 \qquad 4\text{-}3$$

Equations 4-1 and 4-2 are the shallow water momentum equations in two directions, and Equation 4-3 is the equation of mass. The first term on the right-hand side of Equation 4-1 and 4-2 represents the forcing from the wind stress, the second term is the bottom friction, the third term is the pressure gradient and the last term is the wave induced radiation stress forcing. Forcing terms for wind are formulated as:

$$\tau_{wx} = \rho_a C_d |W| W_x \qquad 4\text{-}4$$

$$\tau_{wy} = \rho_a C_d |W| W_y \qquad 4\text{-}5$$

, where τ_w is wind stress, ρ_a is density of air, C_d is the wind drag the coefficient, and W is the wind velocity. In Equation 4-1 until 4-5, x is cross-shore orientation (positive onshore) and y is alongshore orientation (positive to the north). The radiation stresses are obtained from the wave action equation for the short waves. See Roelvink *et al.*, 2009 for details on these equations.

An XBeach model was created for Egmond aan Zee covering 2000 m in alongshore direction and 1000 m in cross shore direction. The grid has a varying mesh in the cross-shore direction, ranging from ~12 m at the offshore boundary to ~5 m near the shoreline. For the alongshore direction, a constant grid size of 10 meters is used. The model is run from rest and begins on 21 August and ending on 27 August 2011 in a continuous run, which covers the entire period of the field experiment. The first 24 hours of the simulation are dedicated to spinning up the model, and therefore will not be used in the analysis. The model version, free parameters and constants used in the model are summarized in Appendix A. The term 'base case' is used from here on to refer to this model setting.

A tidal water level is prescribed at each of the lateral (southern and northern) boundaries with a certain phase lag, so that the water level gradient generates a tidal current (Roelvink and Walstra, 2004). The phase lag is computed through a linear interpolation between two measurement points at IJmuiden (15 km to the South of the model) and Den Helder (43 km to the north). The interpolation yields a phase lag of 1.62 degree (approximately 2 minutes) between the southern and northern boundary, which means an alongshore water level difference of 0.02 meter on average.

Wave data are available every 20 minutes (sea state) from the directional wave-rider buoy offshore of IJmuiden (30 m depth, IJM-1 in Figure 4-1) are transformed using a SWAN model (Booij *et al.*, 1999) to the 10 m depth contour where the offshore boundary is located. The wave data are schematized in such time and alongshore varying boundary conditions of the wave energy and associated bound long wave are generated (Van Dongeren *et al.*, 2003).

To account for flow forcing wind stress, the model is forced by non-stationary wind data, obtained from the wind station at IJmuiden (IJM-2 in Figure 4-1) and forced as a time series of wind speed and direction every 20 minutes.

Figure 4-1: Google Earth view of the study location Egmond aan Zee. The directional wave buoy IJmuiden is labelled as IJM-1 and the tidal gauge and wind station as IJM-2. The insert shows the domain of the numerical model grid.

4.2.2 Data

Egmond aan Zee, a busy beach resort during the summer season, is located in the middle of the uninterrupted part of the Dutch coast between the IJmuiden (port of Amsterdam) harbour moles and the Wadden Sea on the north (Figure 4-1). The morphology is characterized by a double sand bar with reset events for the outer bar on a time scale of ~5 years (Walstra *et al.*, 2012). Wave conditions are mild with 1.3 m significant wave height and 6.4 seconds wave period respectively (Wijnberg, 2002), with significant seasonal variations between summer and winter, with winter waves are higher. The beach is meso-tidal, with a tidal range of ~ 1.4 m – 1.7 m (Wiersma and Van Alphen, 1987). The tidal current is a pronounced hydrodynamic feature at this site, where strong alongshore (tidal) velocity flows to the north during flood tide and to the south during ebb tide.

Data are obtained from the SEAREX field campaign conducted from 22 to 26 August 2011 (Winter *et al.*, 2014). In Figure 4-2, water levels, and waves, and the wind conditions during the experimental period are presented. During the experiment, wave conditions recorded at IJmuiden were mild to calm with significant wave heights varying between 0.3 and 1.2 m and with peak wave periods from 3 to 7 seconds. Incoming wave directions were on average within ± 30° from shore normal. On the last day (26 August) only, wave directions varied from 100° to 170° relative to shore normal (south-easterly waves, *i.e.* radiating away from the field domain). The wind speed varied between calm conditions of 1.5 and ~ 9 m/s with directions from the southwest and occasionally from the north. An exception again was the last day when wind speeds were relatively high for the whole day ranging from 5 to 13 m/s with a persistent direction from the southeast. The grey-shaded intervals in Figure 4-2 indicate the experiment sessions. During each session, four to five drifter deployments were conducted (see also Table 4-1 and Table 4-2).

The bathymetry was surveyed during the experiment with a jet ski mounted single beam echo-sounder and RTK-GPS system (Van Son *et al.*, 2009). The survey covered 2000 m alongshore and 1000 m in the cross shore direction. Bed levels at the offshore end of the domain are about -10 m on average, and about +2.5 m around the shoreward-end of the intertidal part (elevation relative to NAP, Dutch National Datum which is about mean sea level). In Figure 4-3a, the bathymetry data are visualized, showing the typical topography of the Egmond surf zone with two sand bars. The outer bar (offshore-ward) is a sub-tidal bar with a mean depth of 3 m and the inner bar occasionally emerges and shows more variability alongshore. Rip channels are clearly present in the bathymetry data, where the inner bar with mean depth of 0.8 m is interrupted by channels with mean depths of 1.6 m (dark arrows). The merged and rectified time exposure image from Argus cameras (Holman and Stanley, 2007) at the Jan van Speijk Lighthouse confirms the alongshore variability of wave breaking on 22 August 2011 (Figure 4-3b). The wave-breaking signal is clearly seen over the bar as well as in the channels where the wave breaking signals are less pronounced. The rip channels were persistent because morphological change was slow during the experiment.

In the experiment, deployments of GPS-tracked drifters were made to measure the mean currents in a Lagrangian framework (MacMahan *et al.*, 2009). Experiments were focused on the two rip channels (dark arrows in Figure 4-3a) by releasing the drifters in the vicinity of the channels. Data from short duration deployments with low velocities were dropped from the analysis. This includes all drifter data from the second session on day 4 (25 August). After data processing and filtering, 162 drifter paths were obtained from the experiment and will be used in the analysis. A full description of drifter deployments executed during the field survey can be found in Winter *et al.*, 2014.

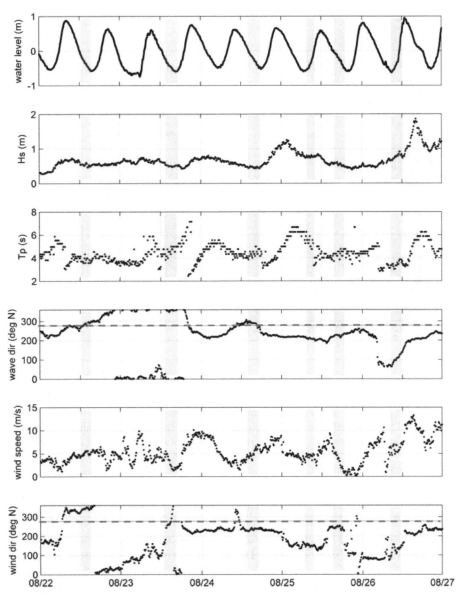

Figure 4-2: Tide, wave, and wind conditions for the period of the study. The tide is interpolated from station IJmuiden and Den Helder. Wave and wind data were measured at the station IJmuiden (Figure 4-1). Grey shading indicates drifter experiment session, and the dashed blue line is shore normal direction

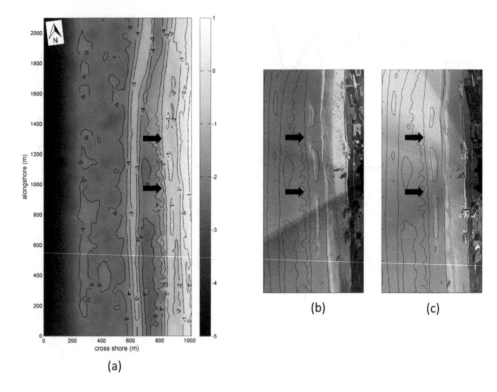

Figure 4-3: Bathymetry data from the jet ski survey (a), the rip channels are indicated by dark arrows (channel 1 upper arrow, channel 2 lower arrow), the shore line is on the right. (b) Zoomed and rectified time exposure image of the site from the Argus cameras at the Jan van Speijk lighthouse on 22nd August 2011, 12.00 GMT and (c) 26th August 2011, 09.00 GMT

4.3 Results

In this section, the results of the model validation are presented. To begin, the simulated tidal current will be compared with previous studies. For the drifter flow paths, virtual drifters in the model were released at the same location and time as in the field experiment and advected with the Lagrangian flow. Simulated drifter trajectories were compared with data in terms of root mean square (RMS) error measures. Finally, the initiation and duration of the rip current intensity, as well as the potential hazard because of the unsteadiness of the rip currents are presented.

4.3.1 Tidal currents

Alongshore tidal currents are prominent hydrodynamic features of the Dutch coast (Short, 1992). Figure 4-4 presents the computed time series of the water level (black) and the alongshore component of velocity (red) obtained at a point at the offshore boundary ($x, y = 0, 1300$ m). The result shows the typical characteristics of a tidal current on the Dutch coast with a tidal range on the order of ~1.5 m. A time lag of ~2 hours between tidal water level and the current attributable to bottom friction is clearly present. The tidal current velocity during flood tide is ~0.7 m/s, with lower values during the ebb tide. This result confirms previous studies from Wiersma and Van Alphen, 1987, and Short, 1992.

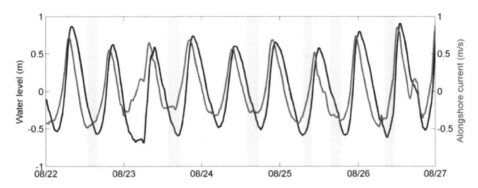

Figure 4-4: Model results for tidal elevation (black, left axis) and the tidal current (red, right axis) at the offshore boundary point at y = 1300 m. Field experiment sessions are indicated by the grey shadings as in Figure 4-2.

4.3.2 Drifter flow path comparison with data

Drifters were seeded in the numerical model simulations at the identical positions and times and allowed them to drift for the same duration as in the field experiment. Hence, a consistent comparison can be made between model output and data. The observations mostly show semi-circular flow patterns with some exceptions. In the following, the model data comparison for several representative flow patterns will be discussed.

Figure 4-5 (first and second row) presents the comparison between observed (Figure 4-5 a-1 through Figure 4-5 d-1) and modelled drifter paths (Figure 4-5 a-2 through Figure 4-5 d-2) for four deployments with varying flow patterns. The bottom two rows are the results from variations of the model and will be discussed later. The model predicts the field drifter trajectories reasonably well. During the first case (22 August, deployment 4), the significant wave height observed at the buoy was 0.48 m with an incoming wave direction from the northwest. Tide was falling and the alongshore tidal current was 0.4 m/s to the south. The wind speed was ~5 m/s from the northwest. Semi-enclosed circulation cells were observed (Figure 4-5-a1) and predicted by the model (Figure 4-5-a-2). Furthermore, a drifter that was released further north than the others moved onshore at the beginning and then floated offshore through the rip channel where it was advected alongshore by the tidal current. This particular pattern was predicted by the model as well.

In the second case shown (22 August, deployment 1), the significant wave height was 0.5 m with wave directions coming from the northwest. The tidal elevation

49

was NAP -0.33 m with tidal current velocities of 0.5 m/s flowing to the south. The drifters' flow paths (Figure 4-5-b-1) show a prominent offshore trajectory through the rip channel followed by an alongshore advection by the current, which the model can predict very well (Figure 4-5-b-2).

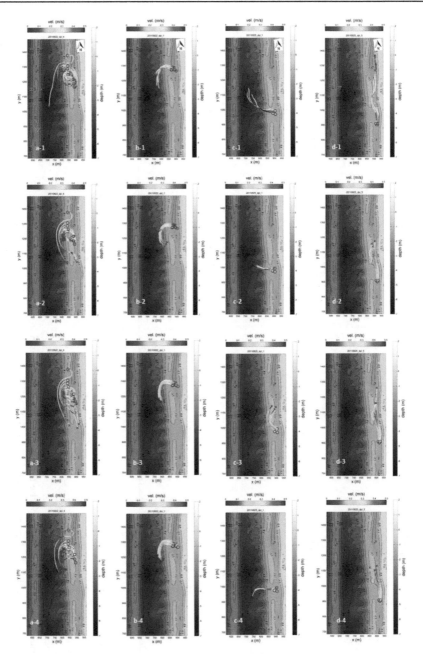

Figure 4-5: Drifter trajectories for four different deployments and varying flow patterns (a: 22nd Aug Deployment 4; b: 22nd Aug Deployment 1; c: 25th Aug Deployment 1; d: 25th Aug Deployment 5). Top row: field data, second row: model results using wave group forcing and wind stress forcing, third row: stationary wave forcing and wind forcing, bottom row: wave group forcing without wind stress forcing. The horizontal colour map indicates drifter speeds. The vertical colour map is the bed level; with dark lines in the background are the bed level contours. Arrows in the top row illustrate the forcing (T: tidal current, H: wave, W: wind).

51

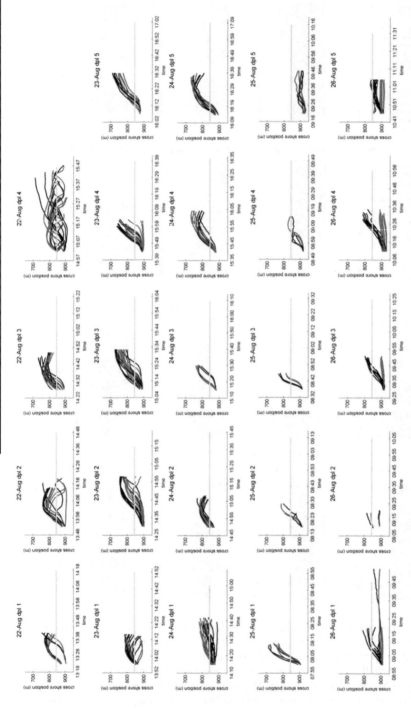

Figure 4-6: Time series of cross shore position of the drifters for Day 1 until Day 5 (top row to bottom row) for 4 to 5 consecutive deployments each day. Black: data, red: model. The grey line indicates the location of the bar-channel position.

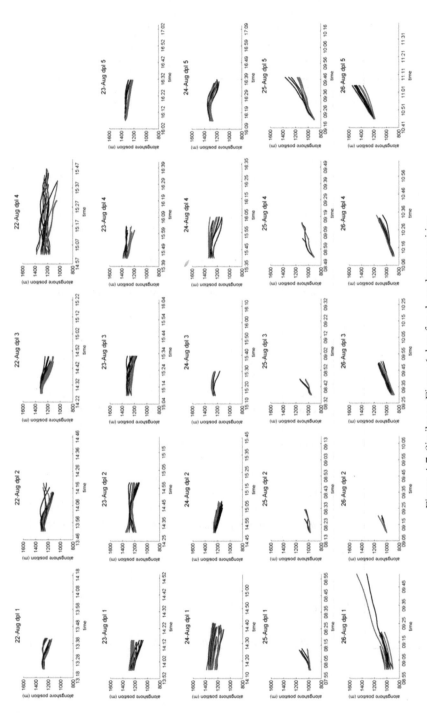

Figure 4-7: Similar to Figure 6, but for alongshore position.

Table 4-1: Relative error $\hat{e}$ of drifter position between data and model results. X is cross shore position, Y is alongshore position. Nd is number of drifter. Plus sign means relative error is higher than base case error, and minus sign means error is lower than base case. Bold means averaged $\hat{e}$ per session.

Session		Base case		Stationary wave		No wind stress	
		$\hat{e}_x$	$\hat{e}_y$	$\hat{e}_x$	$\hat{e}_y$	$\hat{e}_x$	$\hat{e}_y$
Day 1 (22nd Aug)	Nd						
Depl. 1	7	0.08	0.10	0.07-	0.08-	0.05-	0.07-
Depl. 2	8	0.13	0.12	0.13	0.12	0.06-	0.07-
Depl. 3	9	0.08	0.24	0.09+	0.24	0.07-	0.12-
Depl. 4	9	0.17	0.22	0.17	0.24+	0.15-	0.15-
		0.11	**0.17**	**0.12+**	**0.17**	**0.08**	**0.10**
Day 2 (23rd Aug)							
Depl. 1	9	0.14	0.36	0.09-	0.18-	0.13-	0.26-
Depl. 2	8	0.13	0.14	0.20+	0.18+	0.11-	0.14
Depl. 3	9	0.15	0.26	0.21+	0.42+	0.13-	0.18-
Depl. 4	9	0.24	0.36	0.28+	0.40+	0.23-	0.26-
Depl. 5	5	0.14	0.17	0.19+	0.32+	0.11-	0.11-
		0.16	**0.26**	**0.19+**	**0.30+**	**0.14**	**0.19**
Day 3 (24th Aug)							
Depl. 1	10	0.23	0.11	0.36+	0.27+	0.23	0.21+
Depl. 2	12	0.05	0.20	0.13+	0.23+	0.06+	0.14-
Depl. 3	3	0.28	0.20	0.31+	0.20	0.24-	0.16-
Depl. 4	6	0.13	0.26	0.14+	0.17-	0.10-	0.15-
Depl. 5	7	0.06	0.09	0.05-	0.12+	0.06	0.16+
		0.15	**0.17**	**0.20+**	**0.20+**	**0.14**	**0.16**
Day 4 (25th Aug)							
Depl. 1	3	0.13	0.20	0.37+	0.06-	0.10-	0.30+
Depl. 2	2	0.20	0.20	0.17-	0.15-	0.18-	0.27+
Depl. 3	3	0.18	0.19	0.30+	0.14-	0.17-	0.23+
Depl. 4	3	0.12	0.20	0.14+	0.15-	0.12	0.23+
Depl. 5	3	0.04	0.17	0.03-	0.14-	0.05+	0.21+
		0.13	**0.19**	**0.20+**	**0.13**	**0.13**	**0.25+**
Day 5 (26th Aug)							
Depl. 1	8	0.26	0.19	0.28+	0.41+	0.26	0.46+
Depl. 2	2	0.11	0.30	0.15+	0.21-	0.14+	0.27-
Depl. 3	9	0.18	0.13	0.19+	0.21+	0.20+	0.31+
Depl. 4	8	0.22	0.08	0.24+	0.10+	0.24+	0.12+
Depl. 5	10	0.11	0.17	0.10-	0.19+	0.10-	0.29+
		0.18	**0.18**	**0.19+**	**0.23+**	**0.19+**	**0.29+**

Table 4-2: Summary of average peak drifter velocity in the rip neck for each deployment. Bold numbers are the average velocity per session.

Deployment	Average peak drifter velocity in the rip neck (m/s)			
	Data	Base case	Stationary wave	No wind stress
Day 1 (22nd Aug)				
Depl. 1	0.30	0.31	0.31	0.31
Depl. 2	0.27	0.32	0.29	0.31
Depl. 3	0.31	0.32	0.33	0.31
Depl. 4	0.25	0.26	0.29	0.31
	0.28	**0.30**	**0.30**	**0.31**
Day 2 (23rd Aug)				
Depl. 1	0.25	0.12	0.22	0.16
Depl. 2	0.24	0.28	0.20	0.25
Depl. 3	0.25	0.25	0.23	0.24
Depl. 4	0.29	0.34	0.29	0.31
Depl. 5	0.24	0.24	0.21	0.28
	0.25	**0.25**	**0.23**	**0.25**
Day 3 (24th Aug)				
Depl. 1	0.13	0.12	0.16	0.12
Depl. 2	0.22	0.21	0.29	0.20
Depl. 3	0.27	0.30	0.30	0.28
Depl. 4	0.22	0.24	0.28	0.25
Depl. 5	0.29	0.26	0.31	0.28
	0.23	**0.23**	**0.27**	**0.23**
Day 4 (25th Aug)				
Depl. 1	0.52	0.44	0.35	0.43
Depl. 2	0.35	0.51	-	0.44
Depl. 3	0.58	0.51	-	0.47
Depl.4	0.32	0.29	-	0.31
Depl. 5	-	0.21	-	0.20
	0.44	**0.44**	**0.35**	**0.41**
Day 5 (26th Aug)				
Depl. 1	0.19	-	-	-
Depl. 2	0.14	-	-	-
Depl. 3	0.24	-	-	-
Depl.4	0.28	-	-	-
Depl. 5	-	-	-	-

Similarly, Figure 4-5-c-1 presents a strong offshore flow, which the model predicts fairly well, but outside the surf zone the deflection of the flow alongshore was less pronounced (Figure 4-5-c-2). During this deployment (25 August, deployment 1), the wave height was 0.4 m from the southwest. The tidal current was relatively low (0.15 m/s to the south, close to slack water), and wind speed was 4.5 m/s from the southeast.

Figure 4-5-d-1 (25 August, deployment 5) presents a case with a dominant alongshore flow with slightly meandering pattern. The waves were 0.45 m and coming from the southwest. Additionally, the tide was rising, with velocities of 0.3 m/s to the north. The model result shows a similar pattern as in the data, with an alongshore meandering flow pattern without offshore rip current flow (Figure 4-5-d-2). However, the alongshore extent of the drifters was underestimated.

To assess the skill of the model on the spatial excursion of the drifters, the cross shore and alongshore component of the drifters' positions were analysed. Figure 4-6 presents time series of the cross shore position of the drifters from all deployments. The model predicts the offshore flow well beyond the inner bar where the rip channel is located (indicated by the grey lines). The offshore extent is on average 100 m from the bar, whereas some drifters only drift 50 m offshore before they float in an alongshore direction (*i.e.* zero cross-shore movement). During the session on day 5 (26 August, bottom row in Figure 4-6), modelled drifters show nearly constant cross shore position over all deployments. This is in contrast with the field data which display offshore flow. This period was characterized by a relatively strong wind from the southeast, and waves coming from the southeast (~100° nautical, recorded at deep-water wave buoy IJmuiden). This implies that wave energy radiated away from the coast, thus resulting in low wave heights at the offshore boundary of the model, making the rip current flows absent. Offshore-directed rip current flow observed in the data may be due to the south easterly wind. This offshore flow could not be predicted well by the model even with wind stress forcing included in the model.

Similar to Figure 4-6, Figure 4-7 presents the alongshore movement of the drifters. The observed alongshore flow is predicted well by the model. Moreover, model results predict alongshore extent of ~200 m showing fairly good agreement with the field data.

The relative error of drifter positions between the model results and observations was computed, both in cross shore and alongshore orientation to evaluate the model skill quantitatively. The relative error is defined as the RMS error between field and simulated drifter position, normalized with the total distance travelled by the particular field drifter:

$$\hat{e}_x = \frac{e_{x,RMS}}{\sum_{i=1}^{N} ds_i} \qquad 4\text{-}6$$

$$\hat{e}_y = \frac{e_{y,RMS}}{\sum\limits_{i=1}^{N} ds_i} \qquad\qquad 4\text{-}7$$

$$e_{\alpha,RMS} = \left[\frac{1}{N} \sum\limits_{i=1}^{N} (\alpha_{mi} - \alpha_{di})^2 \right]^{\frac{1}{2}} \qquad\qquad 4\text{-}8$$

In equation 4-6 and 4-7, x and y represent cross shore and alongshore direction respectively, $\hat{e}$ is the relative error, e is the root mean square error between model and data, ds is distance travelled by the particular observed drifter over a discrete time, i is time, and N is number of time points. In equation 4-8, m and d denote modelled and data respectively, and *alpha* represents the spatial component x or y. Table 4-1 summarizes the relative error of drifter positions between model (base case) and field data presented as the mean relative error for all drifters per deployment. For the cross shore motion, on average, more than 75% of the modelled drifters give a relative error < 0.2. Higher relative errors were found for day 5 where model results show a uniform alongshore flow instead of an offshore flow, as discussed previously. For the alongshore orientation, approximately 76% of the modelled drifters give a relative error < 0.25. On average the relative error is 0.15 for cross shore position and 0.19 for alongshore position showing reasonably good model skill.

4.3.3 Drifter velocity comparison with data

Data collected from the drifter measurements are the GPS coordinates of the drifters' positions. Flow velocities were determined by computing the distance travelled divided by time elapsed ($v = ds/dt$), with $dt = 2$ seconds or the sampling interval. The rip current velocity was then determined by computing average (peak) velocities of the drifters from each deployment within the rip neck (following Winter *et al.*, 2014).

On average, the modelled drifter velocities only deviated from the observed ones by ~14% (see Table 4-2 for each deployment).

4.3.4 Rip current initiation and duration

The rip current initiation and duration will be evaluated based on the cross shore velocity component from the model over the 5 days simulation length. In Figure 4-8b, a rip current time stack is presented, which displays the 5 minute averaged cross-shore velocity as a function of time and the alongshore coordinate for a position along the inner bar crest profile, at x = 870 m. The red colour indicates the offshore (rip) current, and the blue indicates onshore flow. The onshore currents are fairly large because of the shallow depth over the bar. The tide and

57

wave height at the offshore boundary are plotted in the panel above (Figure 4-8a). A strong (inverse) correlation between the tide and the rip current magnitude is apparent. The rip current becomes active as the tide is falling and increases to a maximum at around low tide. Afterwards, as the tide is rising, the rip current magnitude starts to decrease. During the high tide, the rip currents are mostly inactive. These tidally modulated rip current characteristics are consistent with several studies (Sonu, 1972; Aagaard *et al.*, 1997; Brander and Short, 2001; MacMahan *et al.*, 2005; Bruneau *et al.*, 2009; Austin *et al.*, 2010; Austin *et al.*, 2014). A threshold value of the initiation and cessation of the rip current can be presented as a ratio of wave height to water depth of the bar (*e.g.*, Aagaard *et al.*, 1997; Bruneau *et al.*, 2009), see Figure 4-8c. The rip current is active (red dots in Figure 4-8c) above a threshold value of ~0.55. Bruneau *et al.*, 2009 reported a value of 0.35 based on field data obtained from a meso-macrotidal beach on the Aquitanian coast in France. The same value of 0.35 was also reported by Aagaard *et al.*, 1997 at Skallingen, Denmark.

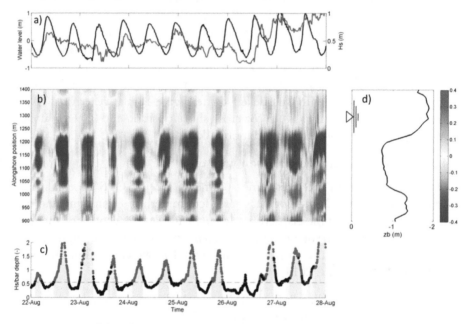

Figure 4-8: (a) Offshore boundary wave height (red) and water level (black), (b) cross shore velocity time stack for alongshore transect at x = 870 m, positive/reddish means offshore directed flow. (c) Ratio of the offshore wave height to the water depth on top of the bar at x,y = 870, 974 m. Grey shades indicate periods when the rip currents are active, and marked by red dots. (d) Alongshore transect of jetski bathymetry at x = 870 m.

In Figure 4-9a, seven low tide time windows of water elevation were averaged. The corresponding cross shore velocity is shown in Figure 4-9b. Time is presented relative to low water in minutes, defined as the time of the local minima from the water level signal. The average value (black line) can be seen as a representative description of condition for each of the low tides in the overall simulation period. The initiation, growth and decay trend of the rip has a similar but inverted shape as the tidal signal. The rip current starts to become active approximately 5 hours before low tide, peaks at low tide, and becomes inactive 3 hours after low tide. Austin *et al.*, 2014 reported for Perranporth Beach, UK, that the rip current velocities start to increase approximately 4 hours before the low tide and peak at 2 hours before and after the low tide. This interval corresponds to maximum wave breaking over the bar region.

Further, the relationship between the rip current intensity and wave forcing was evaluated. Figure 4-10 presents a scatter plot of 5 minute averaged offshore wave heights to bar depth ratio *vs.* dimensionless rip current velocity, where the black circles were obtained from channel 1 and the red crosses from channel 2. Within values of $0.5 < Hs/h_{bar} < 1$, the rip current intensity shows a linear correlation with the dimensionless wave height, which is consistent with field data and laboratory experiments reported previously (*e.g.,* Drønen *et al.*, 2002; Haller *et al.*, 2002; MacMahan *et al.*, 2005). Results from the RIPEX experiment for the open beach of Sandy City, USA (MacMahan *et al.*, 2005) showed that dimensionless wave height and rip intensity correlated fairly significantly, ($r^2 = 0.37$). This value was obtained using data with a wave height to water depth ratio of < 1 (when the waves are not saturated). Using a similar method, $r^2 = 0.32$ was obtained, showing a fairly significant correlation between the wave height and the rip intensity.

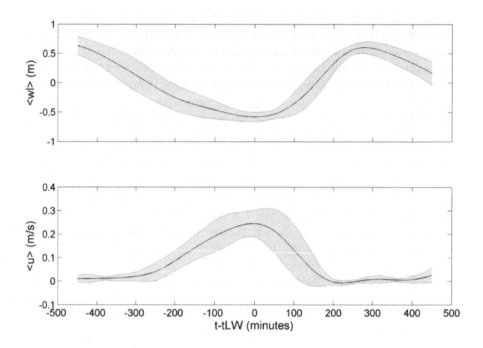

Figure 4-9: Average over seven low tides: (a) water level, (b) cross shore component velocity. Grey shadings are the +/- standard deviation. On the horizontal axis is time relative to Low Water in minutes.

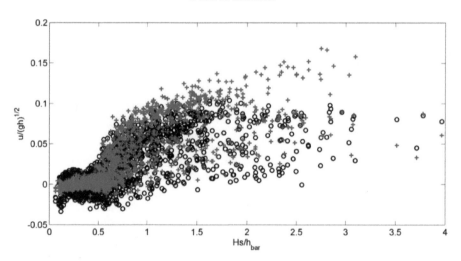

Figure 4-10: Normalized wave heights versus dimensionless cross-shore velocity (5 minutes averaged, positive offshore-directed) from both channels, black circles: channel 1, red crosses: channel 2.

The linear relationship between normalized wave height and the rip intensity implies that the rip current may be active during high tide, but this was not seen in the model results, in that the wave height to water depth ratio never exceeded 0.55 in high water conditions during the experiment. Still, it is possible that rip currents occur for the combination of high water and high waves. To verify this, another simulation with idealized wave forcing and water level was set up. The model was set up with varying wave heights of 1, 1.5, 2, and 2.5 m and corresponding peak wave periods of 5.6, 7, 8, and 9 seconds (enforcing a constant steepness ~ 0.02), with a constant water level of 1 m (high tide tidal state). Each wave condition lasts for 1 hour. With these wave and water level conditions, the offshore wave height ratio to water depth on the bar crest will be consistently above the threshold value of 0.55 aforementioned. From the two rip channels analysed, the results from channel 2 show a monotonic relation between the wave heights and the rip current's intensity during the high tide (Figure 4-11). Channel 1 shows no trend, but weak rip currents do occur. A dimensionless rip velocity of ~0.06, which is equivalent to ~0.3 m/s, may occur at channel 2 with an offshore wave height of ~2 m during high tide. However, large waves only occur in the presence of strong winds in this location, and in these conditions, the risk of drowning may also be less because we can expect fewer swimmers on the beach during stormy weather. Furthermore, during high tide the channels where the active rips are located farther away from the dry beach where the beach users are usually located.

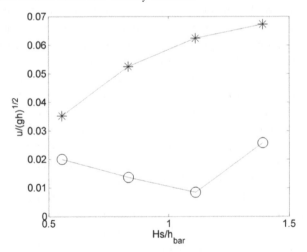

Figure 4-11: Normalized wave heights versus dimensionless cross-shore velocity (circle: channel 1, asterisk: channel 2), for a constant water level of 1 m (high tide state), and varying Hs: 1, 1.5, 2, 2.5 meters.

4.3.5 Rip current circulation and beach safety

The drifter data display a (semi) circular flow pattern (Figure 4-5a), which suggests that the rip currents will transport drifters back to the shore after they float offshore. MacMahan *et al.*, 2010 reported a circulation flow pattern from the field with approximately 7 minutes of cycle time for human drifters at Monterey, California. They reported an infrequent occurrence (10% of the cases) of surf zone exits of the floating GPS drifters, whereas all human drifters were transported back to shore. Miloshis and Stephenson, 2011 performed a human drifter experiment and observed that over 2 days of tests, 3 of 11 human drifters exited the surf zone. Both studies show the possibility that a drifter will be transported back to shore after it has floated offshore with the rip current. Here, the percentage of drifters that are transported back to the shore was quantified. A model run was set up where drifters were released every 5 minutes during low tide (when the rip current is active) with forcing similar to the conditions presented in Figure 4-2 and transformed to the boundary at the 10 m depth contour. Every drifter was allowed to float for 40 minutes (so that the possibility of drifters cycling more than once in the rip cell was minimized). The focus will be on the two channels (Figure 4-3a), and 10 drifters will be seeded shoreward of each channel at approximately equal distances from each other. This scenario was run for seven low tide conditions (low tides between 23 and 26 August). Figure 4-12 presents four snapshots of the drifter positions after 10, 20, 30 and 40 minutes from all seven low tide model simulations. The depth contour of -2 m (beyond which a swimmer cannot stand on the bed anymore) is used as the demarcation line to categorize drifter exits (black line). Leftward arrows and the corresponding numbers indicate the percentage of drifters (relative to the total) that moved from the inside to outside in the 10 minute interval before the snapshot. Similarly, rightward arrows and the numbers indicate drifters that moved back passing the demarcation line towards the shoreline. Based on the model results, at $t = 10$ minutes, most of the drifters (83%) had crossed the -2 m depth contour and moved offshore. After 20 minutes, the number decreased to 4%, with 8% at $t = 40$ minutes. Although some drifters moved from the inside to outside, a larger number moved back onshore in these intervals. Averaged over the 10-minute intervals, the numbers of drifters that were transported onshore are 10%, 12%, and 10% respectively (numbers with rightward arrows in Figure 4-12). Relative to the number of drifters that were at any point in time outside the surf zone, the numbers of drifters that were transported from the outside back inside the surf zone are 12%, 16%, and 14%. This suggests on average, only 14% of the drifters turned back to the shore, in contrast to the two studies previously mentioned. The remaining drifters outside the surf zone would become more dispersed with time, and are mostly advected alongshore depending on the tidal current state and the wave direction. This finding has implications for rip current hazard management: the common guideline for swimmers once caught in a rip current is to swim parallel to the beach. For the Dutch Coast, except during slack water, the tidal current will be dominant outside the surf zone and swimming parallel to the shore will only be the correct strategy when the swimmers swim into the right direction (with the current).

Otherwise the effect can be as dangerous as swimming against the rip. Thus, a rescue strategy requires knowledge of the tidal state. A study on Woolamai Beach and Shelly Beach, Australia suggests that 'do nothing' is an efficient escape strategy and gives higher chances of survival once one is caught in the rip rather than trying to swim parallel to the beach (Miloshis and Stephenson, 2011). Thus, for Egmond aan Zee, a strategy of 'do nothing' and signalling for help is proposed, since the rip currents are unlikely to transport the swimmers back to the shore whilst strong tidal current is likely to be present. In addition, the 'do nothing' strategy can also be combined with swimming back to the shore, because after 'letting go' for e.g. 5 to 10 minutes, one is most probably already outside the offshore flow path of the rip, especially when the rip's orientation is oblique, therefore swimming back onshore can be an efficient strategy.

One of the swimmer safety hazards may be the unsteadiness of the rip currents, resulting in rapidly varying rip current speeds. To this purpose, floating drifter movements' offshore during low tide will be analysed (*i.e.* the time window when the rip current is active, as in Figure 4-8). A model run was set up in which a drifter was released at $x, y = (870, 974)$ m (rip neck), and allowed it to travel for 5 minutes. This drifter release was repeated every 5 minutes throughout the period the rip current was active. Figure 4-13 presents a bar plot of the modelled drifter offshore extent (dx_{off}) for every drifter released in the model. Within 5 minutes, an object can be transported as far as 60 m offshore, even 4 hours before the low tide. As the tide is approaching the low tide, this number increases to ~90 m during the (peak) low tide. Three hours after low tide, when the rip becomes inactive, lower values of dx_{off} are computed.

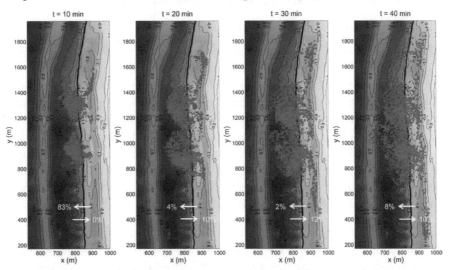

Figure 4-12: Modelled drifter positions (magenta) from seven low tides simulation period, for 4 time instances. Black thick lines indicate -2 m depth contour. The leftward arrow and the corresponding percentage indicate the percentage of drifters that moved offshore. The

63

rightward arrow and the number indicate the percentage of drifters that moved onshore. Green circles are drifter release locations. The percentage is relative to the total number of drifters (n = 3360)

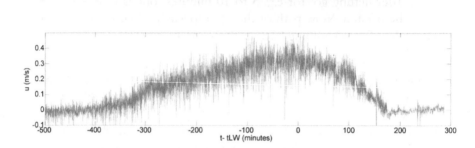

Figure 4-13: Top: Cross shore distance travelled by a drifter within 5 minutes after it has been released in the rip neck. Bottom: Corresponding cross shore velocity (positive offshore). Time is relative to Low Water, in minutes.

4.4 Discussion

In the following sections, the importance of a wave group forcing and wind stress forcing in modelling rip currents will be discussed.

4.4.1 Importance of the wave group forcing

The computational cost between running the model in wave group mode or in stationary mode (short wave resolving computation) can be a factor of five. From an operational point of view, this computational cost may become crucial because the prediction for swimming safety purposes should be performed with reasonable computational and processing time. Here, the investigation will be focused on whether rip currents can still be predicted in terms of the mean flow, initiation, and life span with a stationary wave solver. A model was set up which is forced by regular wave and neglect wave group forcing. For this model, the wave breaking model will be based on Baldock *et al.*, 1998 with a depth-varying breaking coefficient proposed by Ruessink *et al.*, 2001.

Forced by regular waves, the average drifter trajectories showed almost identical results compared with the wave group forced scenario. The drifter flow patterns were consistent with the observations and showed a distinct offshore flow and alongshore drift as well as the semi-circular flow patterns (Figure 4-5-a3, Figure 4-5-b3, and Figure 4-5-d-3) except for 25 August deployment 1 (Figure 4-5-c-3), which will be discussed in the next paragraph. Figure 4-14a presents the modelled cross shore velocity component, collected from channel 2, for the two different wave forcing approaches. The result using stationary wave forcing (red) shows good agreement with wave group forced model result (grey) in terms of tidal modulation of the rip current. This suggests that the rip current occurrence on the time scale of the tide can be adequately predicted by the model using the stationary wave solver. However, the rip current magnitude is under-predicted by the stationary model especially during the period when the rip current is active. Figure 4-14b presents a detail of the cross shore velocity time series of Figure 4-14a, centred around 25 August 2011 08:37 GMT (25 August, deployment 1, Figure 4-5-c1). The five-minute time average of the rip current velocity of the wave group-model result is represented by the blue line. During this specific time period, rip current magnitude is on average 50% lower using the stationary model compared with the wave group forcing model result. On average over the whole low tide from 22 to 26 August, the rip current magnitude is 20% lower using the stationary mode relative to the wave group forcing approach.

Another distinct difference between the two approaches can be found on the 25 August, deployment 1, in which the modelled drifters did not show offshore directed flow using the stationary wave forcing (Figure 4-5 c-3), even though the relative error in alongshore position decreased (Table 4-1). This is the case for all three deployments during that day and is due to the difference in the spatial pattern of the flow field with a different wave resolving approach (see Reniers *et al.*, 2007). Figure 4-15 presents the instantaneous cross shore velocity from the model during drifter release time on 25 August 2011 deployment 3, for the two different wave forcing approaches. Drifter release positions (green crosses) become an important variable affecting the overall Lagrangian flow of the drifters due to the different spatial patterns of the flow field. This is particularly true under oblique incident wave conditions, where both the cross shore and alongshore component forcing interact, and as a result, dictate the offshore extent and orientation of the rip current, as well as the overall spatial pattern of the flow field in the vicinity of the rip channel.

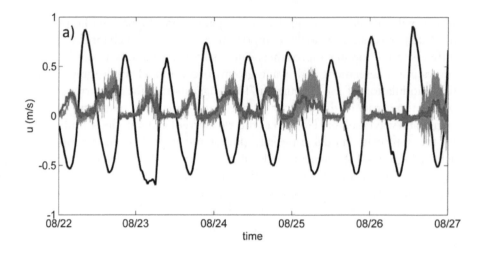

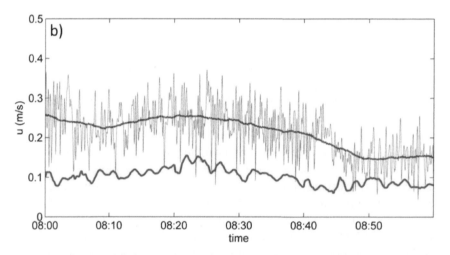

Figure 4-14: a) Modelled cross shore velocity using different wave forcing approaches, collected from a point within the rip neck at $x,y = 870,974$ m, grey: using wave group forcing, red: using stationary wave forcing. Offshore water level is superimposed as the dark thick line. b) A zoomed in time series of a) centred on the third deployment of that day (25 Aug-2011 08:37 GMT, Figure 4-5 c1). The blue line is a 10 minutes running average of the grey cross-shore velocity simulated with wave group forcing.

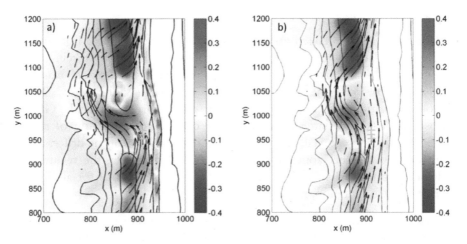

Figure 4-15: Snapshot of cross shore velocity (coloured patch) superimposed by the velocity vector for different wave forcing: a) with wave group forcing, b) with stationary wave forcing, on 25 Aug-2011 08:37 GMT (deployment time of Figure 4-5 c1). Green crosses indicate locations where the drifters were released.

4.4.2 Importance of the wind stress forcing

A simulation with model settings similar to the base case setting but without wind stress forcing was set up. The simulated drifter paths show good agreement with the observed paths (Figure 4-5, bottom row, a-4 until d-4). However, alongshore flow patterns of the drifters become slightly less pronounced. Modelled average peak drifter velocities in the rip channel agree very well with field data (summarized in Table 4-2).

Wind stress forcing as implemented in XBeach mainly affects the alongshore component of the drifter positions, while for cross shore orientation, simulated drifter positions are less sensitive to the wind forcing. For the first and the second day of the experiment, neglecting the wind stress forcing causes a decrease in relative error of the alongshore component of the drifter positions by 40% and 25% respectively (averaged per session, Table 4-1). However, the flow patterns observed in the field were still predicted well by the model. In contrast, during days 4 and day 5, neglecting wind stress forcing increased the alongshore relative error of drifter positions between data and model by 29% and 66%, respectively. During day 4 deployment 1, the alongshore drifter flow showed better agreement with the field data when wind stress forcing was included. For example, without the wind stress forcing, modelled drifters flowed to the south (with the tidal current) after they had floated offshore (Figure 4-5 c-4), whereas the field data indicated northward flow. On day 5, the highest wind speed was recorded, with the dominant directions from the southeast and south. In that case, wind stress forcing improved the model performance.

4.5 Conclusions

Bathymetrically controlled rip current flows were simulated for Egmond aan Zee, The Netherlands, with a model that incorporates wave group forcing, a dynamic alongshore tidal current and wind stress forcing. The model was validated against data from the SEAREX field campaign (Winter *et al.*, 2014). The modelled drifter flow patterns are in good agreement with GPS drifter observations. Flow patterns that resemble offshore directed flow of rip currents are reproduced very well, as well as the alongshore drift of the drifters once they are outside the surf zone. Averaged over all drifter data, the relative error of drifter position between data and model is ~0.15 and 0.19 for alongshore and cross shore orientation, respectively.

Further, the model was utilized to examine rip current initiation and duration. Through numerical experiments, it was found that wave height and water level strongly control the initiation and duration of the rip currents at Egmond. For the period of analysis, the rip currents start to initiate approximately 5 hours before low tide, reach their peak during (peak) low tide, start to decay as the tide is rising, and finally become inactive 3 hours after low tide. Their initiation corresponds to the ratio of offshore wave height and water depth on the bar of ~0.55, which is higher than found on other beaches. Rips may also occur during the high tide; when the offshore wave height to bar depth ratio is > 0.55.

At Egmond, drifters transported offshore by the rip are unlikely to circulate back to the shore (only 14% do so) contrary to observations at other beaches. The drifters rather stay outside the surf zone where they are further advected alongshore by tidal currents.

The common guideline for swimmers once caught in a rip current is to swim parallel to the beach. For the Dutch Coast, except during slack water, the tidal current will be dominant outside the surf zone and swimming parallel to the shore will only be the correct strategy when the swimmers swim into the right direction (with the current). Otherwise, the impact can be as dangerous as swimming against the rip. For Egmond aan Zee, a strategy of 'do nothing' and signalling for help is proposed, since the rip currents are unlikely to transport the swimmers back to the shore whilst strong tidal current is likely to be present. In addition, the 'doing nothing' strategy can also be combined with swimming back to the shore, because after 'letting go' for e.g. 5 to 10 minutes, one is most probably already outside the offshore flow path of the rip, especially when the rip's orientation is oblique, therefore swimming back onshore can be an efficient strategy.

Modelling rip currents with a wave-averaged stationary approach reduces the model predictability on rip current flows. The simplified approach does affect the spatial pattern of the flow field and tends to under-predict the rip current magnitude by 20 % on average.

Wind stress forcing in the model results improves the prediction of alongshore flow, but only during the period when the wind speed is relatively strong.

In the future, more validation for rip current and non-storm (normal) conditions is recommended. In particular, Eulerian velocity data obtained from inside the rip channel would be useful to further validate the model. Moreover, tidal current data near the offshore boundary would also be beneficial as validation material because the tidal current is important for this location. Investigation of inclusion of wind forcing into the model to resolve nearshore (rip) currents is also recommended.

Appendix A. XBeach version and parameter settings

Parameter	Description	Value
Model version	Modified version	XBeach v1.21.3657
gamma	Breaker parameter	0.55
alpha	Dissipation parameter	1.0
gammamax	Maximum ratio of wave height to water depth	2.0
roller	Roller model on/off (1/0)	1
beta	Breaking wave front slope in roller model	0.1
hwci	Minimum depth where wave current interaction takes place	0.1 m
hwcimax	Maximum depth where wave current interaction takes place	3 m
cf	Dimensionless bed roughness coefficient	0.003
nuh	Background horizontal viscosity	0.15 m^2/s
Cd	Wind drag coefficient	0.02

5 Beach bathymetry from video imagery[5]

5.1 Introduction

The importance of having accurate and frequently-updated nearshore bathymetric data is obvious for coastal managers, scientists, and engineers. Especially on a local scale when accurate and high resolution data are needed, conventional techniques can become very expensive because of the necessity of repeated surveys. Therefore, much attention has been paid recently to estimating bathymetry using remote sensing technology.

Water depth in general can be observed remotely using both direct and indirect approaches. Direct remote observation of water depth from remote sensing is simply based on the ability of magnetic waves to penetrate through different layer into the water column. Thus, this technique depends on optical clarity of sea waters. An empirical algorithm then will be used to obtain the actual water depth information. The method is implemented in several technologies e.g. airborne (Benny and Dawson, 1983; Bierwirth *et al.*, 1993; Sandidge and Holyer, 1998; Lee *et al.*, 2001; Adler-Golden *et al.*, 2005), and light detection and ranging (LIDAR) technology (Irish and Lillycrop, 1999). In general, these direct extractions of nearshore bathymetry from airborne technology conform well with ground truth data. Depth can be surveyed from 0 to 6 m, with root mean square error against ground truth ranging from 0.39 to 0.84 m. LIDAR shows relatively better performance, and is able to measure up to 60 m depth, with an

---

[5] Part of this Chapter is adopted from Sembiring et.al, 2014. Nearshore bathymetry from video and the application to rip current predictions for the Dutch coast. *In*: A.N. Green and J.A.G. Cooper (*eds*), International Coastal Symposium. *Journal of Coastal Research*, Durban, pp. 354-359.

accuracy of about 15 cm. However, all direct methods require very clear water, which renders them useless in the turbid North Sea water and in the surf zone.

On the other hand, remote indirect observations of bathymetry quantify the depth based on water surface characteristics. From what is "seen" at the surface, the water depth will be computed by applying an inversion model towards depth-induced-properties to estimate water depth. For nearshore application, the most common inversion model is through linear dispersion relation which relates the wave period and wave length to the water depth. However, there are methods that rely on other than the dispersion relation, for instance wave breaking based inversion and shorelines detected from video images.

In this chapter, a review on bathymetry estimation techniques using remote sensing, both with and without the linear dispersion relation will be presented. A technique based on detection of shoreline from video image will also be briefly reviewed. After the literature review, two techniques: Beach Wizard (van Dongeren *et al.*, 2008) and cBathy (Holman *et al.*, 2013) will be discussed followed by results and discussion on their application for Egmond aan Zee.

5.2 A review on bathymetry estimation through remote sensing technique

5.2.1 Depth inversion via wave dispersion relationship

Depth inversion using wave dispersion relation relates wave parameters such as wave period T and wave length L, to compute the bathymetry. The surface gravity wave dispersion relation (Dean and Dalrymple, 1991) reads:

$$\omega = \left(gk \, \tanh(kh) \right)^{\frac{1}{2}}$$
5-1

,where ω is angular frequency $(2\pi f)$, f is the wave frequency, g is the acceleration of gravity, k is angular wave number $(2\pi/L)$, L is the wave length, and h is the local water depth. Since water waves become less dispersive as they propagate toward shallower water, the wave celerity becomes more strongly dependent on water depth rather than on the wave parameters; therefore the phase speed, c, has been more practically used for depth inversion by substituting for either one of the frequency and the wave number as shown below:

$$c = \frac{\omega}{k}$$
5-2

$$c = f(\omega, h)$$
5-3

, which can be easily evaluated using an iterative procedure or approximate expressions.

Observing parameters of interest to be used in the linear dispersion relation can be done through various methods, for instance airborne video (Dugan *et al.*, 2001), land video station (Holland and Holman, 1997; Stockdon and Holman, 2000), synthetic aperture radar SAR (Greidanus, 1997), spot satellite images (Leu *et al.*, 1999), and microwave X band radar (Bell, 1999; McNinch, 2007). In the following, a method using video imagery will be briefly discussed.

Dugan *et al.*, 2001 estimate surface currents and water depth using a digital framing camera, by capturing temporal sequences of optical images of shoaling of ocean waves. This technique put information into 3D data: space-space-time data cube. The 3D data cubes are analysed to estimate wave properties of wavelengths and surface currents using 3D fast Fourier transform (FFT) analysis. The technique is able to effectively separate spatial and temporal modulations due to waves. Applying this technique around Army Corps of Engineers Field research Facility (FRF), they reported wave parameters; bathymetry and surface currents obtained from the work are in general within 10% of concurrent field measurements. The 3D cube technique inferred bathymetry by relating wave parameters and water depth through the wave dispersion relation.

Stockdon and Holman, 2000 estimated nearshore bathymetry by computing cross and alongshore wave number from time stacks. Wave frequency is estimated by calculating intensity based energy spectra from a cross-shore time stack at each cross-shore location and averaging spectra over all cross-shore locations. The input frequency will be the spectral peak frequency. For the cross-shore wave numbers, frequency domain complex empirical orthogonal function (CEOF) analysis is applied to the normalized cross-spectral matrix of the pixel intensity. This technique was applied at Duck, NC. The result shows a good agreement with field measurements, with mean relative error was 13% with tendency of overestimation.

Misra *et al.*, 2003 developed an algorithm to estimate water depth using remotely sensed data, based on extended Bousinesq equation (Wei *et al.*, 1995; Chen *et al.*, 2000; Kennedy *et al.*, 2000), which take into account wave breaking and wave induced currents. The work is done for 1D profile synthetic case, where the input is obtained by doing simulations using fully nonlinear Bousinesq equations. They use wave celerity in the inversion model to derive water depth. The wave celerity is calculated based on time lag of surface elevation and/or orbital velocity. For several cases, they found that for monochromatic wave conditions, the agreement between computed depth and actual depth is good. Using a non-linear method of the model, depth prediction is improved by 10% compare to the linear method. For monochromatic waves with presence of currents, they modified the algorithm to take into account mean flow effects, and this modification improved depth prediction by 10% compared to initial settings of the algorithm. However, when they used irregular wave conditions, generated using TMA spectrum with varying peak enhancement factor; they found that the skill of the algorithm decreases as the wave spectrum becomes broad-banded.

Plant *et al.*, 2008 evaluated two approaches in determining ocean wave number, which are power spectral density method and cross spectral correlation. They stated that using power spectral density approach, e.g. Stockdon and Holman, 2000, which requires application of 2D Fast Fourier Transform to full frame images, is probably not efficient enough. In addition, the spatial resolution of the wave number (or wave phase speed equally) is typically in order of 100 times the image pixel resolution. Therefore, they focus on the second method, which is using maximum cross spectral intensity coherence to estimate wave number. They develop a solution based on this approach in sectioning fashion (tomographic analysis), by utilizing a nonlinear based inversion method. One of the advantages of the proposed method is that it provides error predictions, which is very useful for evaluating model design. Moreover, they addressed the spatial resolution of the wave number using the second approach, and they show it can be improved up to ten times compared to the first approach. Finally they delivered a depth inversion scheme based on wave number parameters using the wave dispersion relation. They apply an iterative procedure in computing the local depth. Following up this work, Holman *et al.*, 2013 developed an algorithm, called cBathy, with which estimates of bathymetry can be obtained on an operational daily basis, introducing a running average procedure using Kalman filtering. The cBathy theory and the application will be discussed in Chapter 5.4.

5.2.2 Depth inversion using other methods

Initial work on bathymetry inversion not using wave dispersion relation was introduced by Lippmann and Holman, 1989. They utilize persistent high intensity caused by wave breaking in the breaker zone, in order to obtain statistically stable breaking wave patterns. They did this by exposing video images over 10 minutes period and then averaging them out. The high intensity in the image will be a proxy of wave breaking, thus depict morphology such as sand bar existence beneath.

Aarninkhof *et al.*, 2005 derived a technique in estimating the cross shore profile from video imagery. In principle, the method is also based on wave breaking/dissipation feature captured in the exposing and averaging video. This method derives the profile by using simple linear relationship between "accretion and erosion" and "difference of modelled and measured wave roller dissipation". This work shows that the root mean squared error of the computed bathymetry is approximately 40 cm (in two cross sections considered). Maximum error was observed (up to 80 cm) in the trough region. The reason for this is due to the lacking of wave dissipation information (from the image) in this region (waves do not break on the trough, leads to no intensity in the video image). Further, the previous method was extended to two dimension horizontal using Delft3D model system. This work showed potential ability in predicting accurately complex bathymetry. However, limitations of the method exist. This method requires cognition of parameters which are most of the time unknown. Aarninkhof *et al.*, 2005 categorized the parameters in to three clusters: hydrodynamic cluster, video cluster, and morphological cluster). Moreover, it shows large errors relative to ground truth bathymetry in the bar trough, just like in the 1D model, and due to the same cause, namely lack of intensity information in the bar trough region.

Scott and Mason, 2007 apply data assimilation method to improve morphodynamic model prediction of bathymetry. The morphodynamic model is a 2D model, which update the bathymetry based on sediment transport rate. They use hydrodynamic model developed by Proudman Oceanographic Laboratory, which computed water flow based on tidal condition, and also able to predict surge level. In order to improve prediction skill of the model for bathymetry, they integrate the morphological model with data assimilation scheme. They collected intertidal kriged bathymetry data from sets of water lines obtained from SAR (Synthetic Aperture Radar) images, and used these data as additional information to the integrated model by way of a data assimilation procedure. They found that applying the data assimilation scheme improved the model skill compare to applying morphological model alone. For instance, applying data assimilation has successfully introduced erosion along a certain channel while this erosion was not observed from morphodynamic model result. They concluded that data assimilation in coastal morphology is a promising tool to be considered in the future. In addition, the data assimilation scheme also allows using other sources of observation rather than just intertidal bathymetry/waterlines. However, they suggested three things for future considerations as they found it could be improve the model skill. Firstly, they applied homogeneous observation error (uncertainty in the observed data). They suggested considering using spatially varies uncertainty, for example greater uncertainty in area where data are less trustworthy. Secondly, the procedure they introduced does not update the uncertainty in time and space. They suggest applying sequential update, since the background bathymetry should be in higher accuracy after a certain bathymetry observation is applied. Finally, they indicated that observation and background error may not be uniform and Gaussian.

van Dongeren *et al.*, 2008 furthered previous efforts in deriving bathymetry from video image technique. The main improvement of this research relative to the one carried out by Aarninkhof *et al.*, 2005 is that the assimilation scheme extends the number of sources/remote sensed data used, and reduces the number of free parameters. In their study, they used three kind of sources/remotely sensed data: roller energy dissipation, intertidal bathymetry, and wave celerity. The roller energy dissipations are derived from Argus time exposures image of breaking intensity. On the other hand, the intertidal bathymetry yielded using Intertidal Beach Mapper (IBM, Aarninkhof *et al.*, 2003). For the wave celerity, they used estimation of such parameter from video pixel time series as described in Plant *et al.*, 2008. Compared to the work by Scott and Mason, 2007, van Dongeren *et al.*, 2008 apply sequential update, meaning that background error is varying in time, as we update the bathymetry. More detailed description and the application of this technique will be presented in Chapter 5.3.

5.2.3 Shore line detection from video images

This method focuses on estimating shoreline location and assembles the estimates over many different tidal elevations to obtain (intertidal) bathymetry. There are several methods available to estimate shoreline location from video images: using maximum intensity on the swash zone (Shoreline Intensity Maximum, SLIM method, Plant and Holman, 1997; Madsen and Plant, 2001) , Colour Channel Divergence (CCD) model of Turner *et al.*, 2001, Pixel Intensity Clustering (PIC) model from Aarninkhof *et al.*, 2003, and Artificial Neural Network (ANN) method by Kingston, 2003. The performance of these methods were analysed in Plant *et al.*, 2007, which shows that all these methods shows significant skill in predicting intertidal bathymetry, and can be applied in diverse coastal environments.

In this thesis, the model suggested byAarninkhof *et al.*, 2003, the so-called Intertidal Beach Mapper (IBM) will be used. The intertidal bathymetry is obtained by assembling a set of waterlines from time exposure video image over a tidal cycle. Waterlines are detected by applying automated clustering of the 'wet' part of the beach and the 'dry' part pixels in HSV (Hue-Saturation-Value) colour space, known as PIC method as mentioned in the previous paragraph. This work is furthered by Uunk *et al.*, 2010 to apply user un-supervised shoreline production. This un-supervised method, called Automatic Shoreline Mapper (ASM hereon) is able to deal with intelligent selection of a shoreline within a certain region of interest/search area and unsupervised quality control of the obtained bathymetry data. Uunk *et al.*, 2010 show that applying the method by user-supervised results in root mean squared error of the bathymetry of 0.28 m while 0.34 m using automated user-unsupervised method they suggested. This is a potential procedure to be considered since a lot of man hours can be reduced. The ASM and IBM will be used in to obtained intertidal bathymetry for Egmond aan Zee (later in Chapter 5.5).

5.3 Beach Wizard: Nearshore bathymetry estimation using wave roller dissipation from video

5.3.1 Theory

Beach Wizard (after van Dongeren *et al.*, 2008) is a data assimilation model, and meant to combine measured/observed parameter with background/prior parameter that we had, to predict or compute a new expected value of the parameter. Here, the parameter is bathymetry. Observed bathymetry (h_{obs}) will be derived (indirectly) from remote sensed data (e.g. video image), while the prior bathymetry (h_{prior}) can be from a measurement or a best guest bathymetry. Each of this bathymetry value has its own uncertainty. Therefore, based on optimal least-squares estimator, the new/updated bathymetry can be computed as:

$$h_{update} = h_{prior} + \alpha(h_{obs} - h_{prior})$$

5-4

In equation 5-4, α is optimal weighting factor of the observed and prior bathymetry, representing to what extend we trust observed data. It reads:

$$\alpha = \frac{\sigma^2_{prior}}{\frac{T_s}{\Delta t}\sigma^2_{obs} + \sigma^2_{prior}}$$

5-5

In Equation 5-5, Ts is the simulation duration for a given observed data, and t is the numerical time step in model computation. Once we have the updated parameter (in this case bathymetry), uncertainty of this updated bathymetry can be computed as following:

$$\sigma^2_{update} = \alpha\frac{T_s}{\Delta t}\sigma^2_{obs}$$

5-6

Equation 5-6 will be executed simultaneously, where this updated uncertainty will become prior uncertainty in the next time step.

However, we do not have direct observation of bathymetry (h_{obs}), instead, we have remote sensed observation of time-average image intensity (which can be related to the water depth/bathymetry 'h'). Therefore, an inverse model is used to relate this remote sensed observation with the bathymetry, and using chain rule, we will have:

$$h - h_{obs} = \left(\frac{df}{dh}\right)^{-1}(f - f_{obs})$$

5-7

In Equation 5-7, f is computed intensity, and f_{obs} is observed intensity. Uncertainty of the observed/measured data will be defined as follows:

$$\sigma^2_{obs} = \frac{\varepsilon^2 + (f - f_{obs})^2}{\left(\frac{df}{dh}\right)^2 + \delta^2_i}$$

5-8

In Equation 5-8, σ_{obs} is the measurement error for a certain sources that we use, δ is so called noise level which ensure the denominator will not go to zero when the derivative df/dh is zero.

Finally, the assimilation model of Beach Wizard, using multiple observation sources, can be expressed as follows:

$$h(t + \Delta t) = h(t) - \alpha \sum_{i=1}^{S} \frac{\frac{df_i}{dh}}{\left(\frac{df_i}{dh}\right)^2 + \delta i^2} (f_i - f_{i,obs})$$
5-9

A bed update scheme as presented here considers only local differences of intensity. To account for effects of spatial distribution of the wave evolution from offshore towards onshore, an additional regulation is needed so that the bed changes will take place at offshore area first and then moving nearshore. With this approach building erroneous bathymetry near the shore line as described in van Dongeren *et al.*, 2008 can be mitigated. A variable γ is introduced and applied into Equation 5-9 (Sasso, 2012):

$$\gamma = \cosh\left(\left(\frac{1}{100}\left(x - \frac{0.6t(x_f - x_i)}{T_s} - x_s\right)\right)^{-10}\right)$$
5-10

, where x is cross shore distance, t is time, T_s is simulation length, x_f-x_i is the total length of the x-axis, and x_s is a buffer value in which the factor begins to move onshore. Here, we define x_s as a function of wave dissipation, so that bed change factor starts when wave dissipation reach 5% of the maximum.

5.3.2 Wave dissipation maps from video

Dissipation maps are produced by creating intensity maps of rectified images (images from different cameras view are rectified and merged to have a plan view image). In general, the process of producing a dissipation map from time exposure images can be summarized as follows:

1. Initiate computational grid domain.
2. Transform the domain into Argus coordinates.
3. Produce plan view intensity image (dissipation map) comprises time exposure images from different cameras.
4. Cut off the dissipation map. In this step, we will cut some parts around the beach line to have a correct dissipation patch, instead of using all intensity appear in the image which would include sandy beach areas.
5. Scale up the intensity in such that we have a correct quantitative figure of wave dissipation.

The scale up is done by multiplying the intensity from the image with total dissipation (from actual wave condition), divided by the total intensity. This can be expressed as following (van Dongeren *et al.*, 2008):

$$D(x, y) = \left(\frac{I_p(x, y)}{\iint I_p \, dx \, dy} \right) \int Ec_g \cos \theta \, dy \qquad \text{5-11}$$

In Equation 5-11, I_p is intensity (in this case dissipation) obtained from the camera, D is roller dissipation (in a correct sense of wave energy dissipation figure, unit of energy), and $Ecg\cos(\theta)dy$ describes the incoming wave energy condition from offshore ($E = \frac{1}{8}\rho g H_{rms}^2$), c_g is wave group velocity, and θ is wave angle (with respect to shore normal). Images with bad quality (raindrops, over lighted by sun glare, etc.) are skipped. For that purpose, a selection tool is developed, where we can "save" good images and put their epoch time in the image list, while the images with rain drops, over lighted by sun glare, etc., will be skipped.

The Beach Wizard algorithm is implemented as a sub-module in XBeach program structure. Computed wave dissipation will be provided by the short wave module in XBeach and later will be used for assimilation with the scaled video intensity data. The workflow diagram of beach Wizard within XBeach can be seen in Figure 5-1.

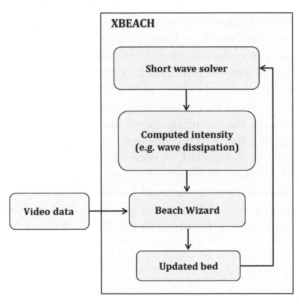

Figure 5-1: Flow diagram of Beach Wizard implementation in XBeach

5.3.3 Application (August 2011 field data)

Wave dissipation maps were produced, using images starting from 5 August 2011 until 22 August 2011. A manual selection was carried out so that bad images due to for example rain drops and sun glare were skipped. At the end, 37 dissipation maps were obtained and will be used in the test. The dissipation map obtained from the video shows the existence of rip channels. In Figure 5-2, two dissipation maps from video imagery are presented, taken from 18 August 2011 GMT11.00 and 09 August 2011 GMT06.30 as examples. The alongshore white band is clearly interrupted by two channels (Figure 5-2, left), which correspond to the alongshore trough seen in the surveyed bathymetry (superimposed as contour line). The existence of the outer bar is seen from dissipation maps during relatively higher wave height (right).

Using an alongshore uniform topography as the initial bathymetry input, Beach Wizard evolves the bathymetry towards the ground truth. The progression is depicted in Figure 5-3. At the beginning, computed dissipation shows mostly alongshore uniform pattern representing the underlying bathymetry. Moreover transects of the dissipation map also show a clear difference in location of the peak dissipation (location of the bar crest). The dissipation map then becomes more similar to the ground truth as number of images used increases.

Figure 5-4 shows the results of Beach Wizard application at the end of the simulation. Estimated bathymetry is displayed on the left, the surveyed bathymetry is in the middle, and the difference between the two is on the right. In general, the bathymetric features are produced very well by Beach Wizard. The outer sand bar, which is located around $x = 400$ m is reproduced. In addition, features near the shoreline, where the inner sand bar typically disrupted by channels is also very well predicted. Two channels and three berm features appear clearly in the bathymetry estimate as they appear in the surveyed bathymetry. Cross shore transects of the bathymetry are presented in Figure 5-5. Beach Wizard was able to nicely move the outer bar location into the right direction. However, the bar-crest position is shifted too much onshore, and the trough depth is underestimated. In contrast, alongshore bar-crest profiles along the inner sand bar show that the bathymetry from Beach Wizard can accurately produce the features in the surveyed bathymetry where the rip channels are clearly observed at $y = -400$ m and $y = -50$ m (Figure 5-5, right panel). The global root mean square error of the Beach Wizard estimate is 0.80 m.

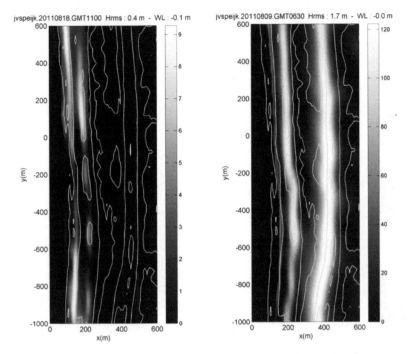

Figure 5-2: Video intensity map of wave dissipation for two different time instances and wave/water level condition. Contour lines are surveyed bathymetry. Signal of bar-channel morphology clearly seen from the white band (left), as well as the outer bar (right).

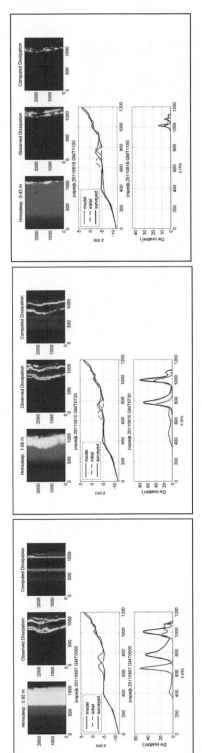

Figure 5-3: Beach Wizard computation progress. Left: after 2 dissipation maps, middle: after 18 maps, right: after 26 maps. For each sub-figure, top panel from left to right: Hrms, dissipation map from video, modelled dissipation; middle panel: profiles of bathymetry; bottom panel: profiles of dissipation.

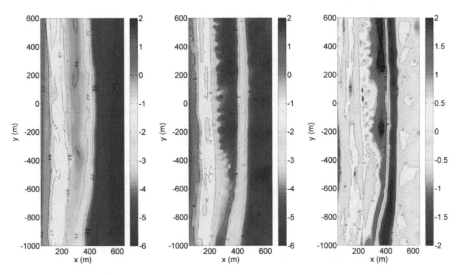

Figure 5-4: Bathymetry estimate from video using Beach Wizard (left), surveyed bathy (middle), and the difference (right). Difference is video minus surveyed.

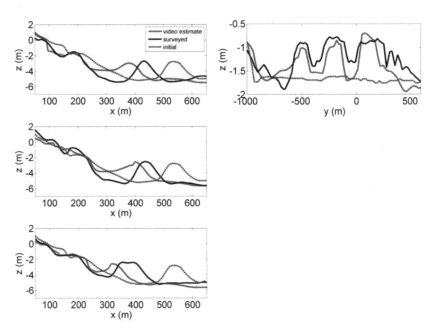

Figure 5-5: Profiles of different bathymetry sources. Left: from top to down profile at y = 500, -200, and -900 meter. Right: alongshore bar crest profile.

5.3.4 Discussion

Bathymetry estimate obtained using Beach Wizard – dissipation maps resemble rip channel features observed in surveyed data. Beach Wizard can accurately produce the features in the surveyed bathymetry where two rip channels are clearly observed between y ~ -400 m and y ~ -50 m. However outer bar-crest is shifted too much onshore and the trough depth is underestimated. The first issue is related to the offset between the maximum intensity from the time exposure and the actual bar crest position (Morris, 2013). The time exposure image is an average over a 10 minutes period. As signal will also strong not only at the location where the bar crest is located but also at the landward part of it due to the persistent foam generated by wave breaking, the 'maximum' location then shifted landward due to this averaging. The later issue, which is the underestimation of the depth on the trough, is also the direct consequence of the persistent foam. The model interprets this signal as a part of the energy dissipation, therefore will result in shallower estimates.

5.4 cBathy: Nearshore bathymetry estimation using pixel intensity time stacks

5.4.1 Theory

cBathy (Holman *et al.*, 2013) is an algorithm which uses time stacks of Argus video images to obtain the bathymetry. The video pixel intensity, $I_p(x, y, t)$, are collected from the images in two dimension spatial domain, as illustrated in Figure 5-6. The algorithm consists of three steps. The first step involves mainly signal processing and data analysis technique in order to obtain depth estimates corresponding to several frequency bands. In step two, these multiple estimates will be merged to obtain one single estimate using many weighting factors. And finally in step three, the single estimates will be smoothed by applying Kalman filter (Kalman, 1960) to obtain a smooth running averaged estimate. Each step will be elaborated in the following which is largely summarized from Holman *et al.*, 2013.

Step 1 Frequency-dependent analysis

First, the full data set of the video pixel intensity will be transformed from time domain into frequency domain using Fast Fourier Transform technique (FFT) as described below:

$$I_p(x, y, t) \rightarrow G(x, y, f) \qquad \qquad 5\text{-}12$$

, in which G is the complex variable comprising Fourier coefficients, the arrow represents FFT, and f is frequencies. As the method resolves phase, G will be normalized so that the resultant of the Fourier coefficients will be unity. This full set of the data then will be sub-sampled into smaller regions (represented as green dots in Figure 5-7). The sub-sampled data will be processed to extract the cross spectral matrix out of it:

$$C(x,y,f) = G(x_i, y_i, f) \cdot G(x_j, y_j, f)$$ 5-13

, in which C is the cross spectral matrix for frequency f, and i and j represent row-column orientation of the data. Note that the frequency f is actually within a certain band, so the variables in Equation 5-13 can be seen as an average over the frequency band. The cross spectral matrix is obtained for many frequency bands. For efficiency purpose, only frequencies with the largest total coherence $\sum|C|$ will be considered for the next process as the rest of frequency provide no useful signals. The number of frequency bands used is a user defined value, and for this application we use four. After four-most coherent frequency are obtained, for each frequency, the dominant eigenvector and the associated eigenvalue will be computed from the C (i.e. extracting the most dominant motion from the signal), which will result in phase structure map $v(x,y,f)$. The wave number then will be computed based on nonlinear fitting so that an optimum wave number will be obtained that gives the best match between observed phase structure from the video signal and modelled one on a forward model as follows:

$$v' = \tan^{-1}\left(\frac{imag(v)}{real(v)}\right) = \exp\left(i\left[k\cos(\alpha)x_p + k\sin(\alpha)y_p + \phi\right]\right)$$ 5-14

, in which v is the phase structure (from the video data, it is the dominant eigenvector $v(x,y,f)$, k is wave number, α is wave direction, and x_p and y_p indicate the pixel points within the tile. Depth estimate $\tilde{h}$ can then be obtained by applying the linear dispersion relation (Equation 5-1) using derived wave number k and the frequency f. For each depth estimate, 95% confidence intervals are also computed. An illustration showing the process from all possible frequencies data to become frequency dependent depth estimates is presented in Figure 5-8.

Depth estimation at this step is performed for every x_a, y_a point, and the analysis tile will be centred around this point, with size $[x_a \pm L_x, y_a \pm L_y]$ (see Figure 5-7). The size of the tile is linearly increasing towards offshore direction. This is useful as the natural variability of the bathymetry is taken into account, for instance near to the shoreline length scale of bathymetric features are small, whereas far offshore bathymetric features length scale may become very large (minimum variability). A constant value of κ is used as a scale factor, so that at the outer offshore domain, the size of the analysis tile will become $x_a \pm \kappa L_x$, $y_a \pm \kappa L_y$ (see Figure 5-7).

Some quality controls are introduced in order to avoid returned unrealistic values. Several parameters obtained from the previous process are used. The skill of the nonlinear fit, s, is used in which depth estimate which gives s less than 0.5 will be rejected. Moreover, the (normalized) eigenvalue, λ, has to be greater than 10.

Finally, depth estimates $\tilde{h}$ which are outside maximum/minimum user defined value are also not accepted. See Table 5-1 for parameters and the values used in cBathy application for Egmond.

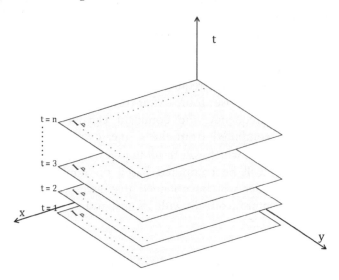

Figure 5-6: Illustration of pixel intensity (I_p) stack in space and time

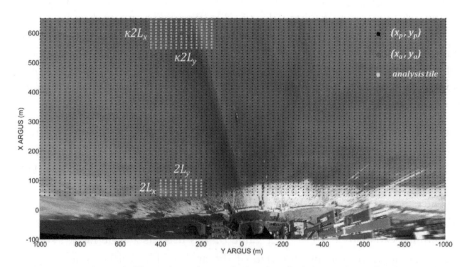

Figure 5-7: Schematic illustration of array design for pixel time series collection (black dots) and the running analysis tile in cBathy (green dots). See text and Table 5-1 for symbols' explanation. Number of points shown was reduced for clarity.

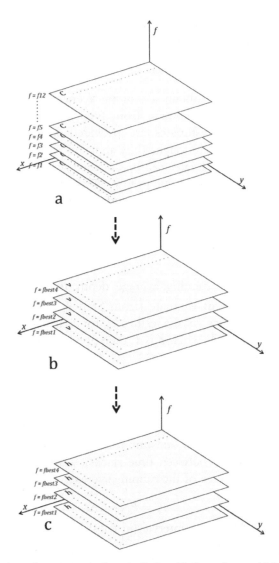

Figure 5-8: Illustration of processes in Step 1 cBathy. (a) Cross Spectral Matrix for all possible frequencies (f_b in Table 5-1), (b) eigenvector for 4-most coherent frequencies *fbest*, and (c) frequency dependent depth estimates. These processes are sub-domain (tile) based analysis.

Step 2 Frequency-independent depth estimation

The aim of Step 2 is to obtain one single depth estimate, out of frequency-dependent depths from Step 1. The single depth will be estimated by best weighted fit of the linear dispersion relation to the f-k pairs from Step 1. The weighting factor is in a function of: the skill, s, the eigenvalue, λ, and a Hanning filter Γ, which regulates value of a coefficient as a function of distance (e.g.

distance from each pixel point within the tile to the analysis point x_a, y_a). The weighting factor is a product of the three aforementioned, $w2 = s\lambda\Gamma$.

Step 3 Running average depth estimation

The aim of Step 3 is to obtain a running average estimate of bathymetry by combining the individual estimate from Step 2 with the previous (in time) running average. Since estimates are provided with confidence (error), the estimates can be objectively weighted by applying Kalman filter (Kalman, 1960). By doing this, smooth estimates can be obtained since individual estimates from Step 2 may include many gaps due to low signal or bad image quality due to sun glare and rain drops. The depth will be updated in the Kalman filter as follow:

$$\overline{h}_k = \overline{h}_{k-1} + K\left(\hat{h}_k - \overline{h}_{k-1}\right)$$

5-15

in which $\overline{h}_k$ is current running average depth, $\overline{h}_{k-1}$ is the previous running average, $\hat{h}_k$ is the current estimate (from Step 2), and K is the Kalman gain, which entails confidence in the estimate, for instance if K is zero, the new estimate is not used, while when K is 1, then previous running average is ignored. The K is computed as follow:

$$K_k = \frac{P_k^-}{P_k^- + R}$$

5-16

, in which P_k^- means the error variance (σ^2) of the running average, and R is the variance of the estimate. To account for decrease of confidence due to un-modelled period (period between time stacks), a so called process error (Q) is added to the error variance of the running average,

$$P_k^- = P_{k-1} + Q\Delta t$$

5-17

In Equation 5-17, Δt is time interval (in days) between k and $k-1$. The updated error variance then computed as:

$$P_k = (1 - K_k)P_k^-$$

5-18

The process error Q, is modelled as a Gaussian shaped line as a function of cross shore position x and wave height:

$$Q(x, H_{m0}) = C_Q H_{m0}^n \exp\left\{-\left[\frac{(x - x_0)}{\sigma_x}\right]^2\right\}$$

5-19

in which x_0 is the centre of the curve (the location where the most dynamic bed changes is expected), σ_x is the spread of the curve, and C_Q is a constant (see Table 5-1).

The estimates returned by cBathy are water depths; therefore have to be corrected with the actual tide to obtain the correct bathymetry.

5.4.2 cBathy set up and pixel time stack collection for Egmond

Applying the algorithm for wind sea dominated beaches like the Dutch Coast requires some adjustments of the parameters used by Holman *et al.*, 2013. Here, we apply some changes relative to the default settings such that the method is more suitable in wind sea-dominated environments with shorter wave lengths like the Dutch Coast. Pixel intensity collection is designed in an array (black dots in Figure 5-7), with resolution of 3 meters and 10 meters for cross shore and alongshore direction respectively. Analysis points (red dots in Figure 5-7) will be every 10 and 20 meters (in x and y) where L_x and L_y are 20 and 40 meter respectively. 'Deep water' is taken at 8 meter depth. A complete list of values and parameters for analysis and pixel collection are summarized in Table 5-1.

Table 5-1: Parameters for cBathy pixel collection and the analysis for Egmond Jan Van Speijk ARGUS station.

Δx_p	3 m	Pixel cross shore spacing
Δy_p	10 m	Pixel alongshore spacing
F_s	2 Hz	Sampling frequency
t_p	1024 sec	Pixel time series length
Δx_a	10 m	Cross shore analysis point spacing
Δy_a	20 m	Alongshore analysis point spacing
L_x	20 m	Analysis tile dimension in x
L_y	40 m	Analysis tile dimension in y
$\varkappa$	5	Factor scale for tile size
h_{min}	0.25 m	Minimum acceptable depth
h_{max}	8 m	Maximum acceptable depth
s_{min}	0.5	Minimum acceptable skill
λ_{min}	10	Minimum acceptable normalized eigenvalue
f_b	$\left[\frac{1}{15} : \frac{1}{50} : \frac{1}{3}\right]$	Frequency bins
N_{keep}	4	Number of frequency bins to retain
C_Q	0.04/day	Constant in process error curve
σ_x	75 m	Spread in process error curve

5.4.3 Application (June 2013 field data)

The cBathy algorithm is tested for Egmond aan Zee using pixel time stacks data collected from 29 March 2013 until 7 June 2013, comprises 218 different time stamps of pixel stack data. From March until May, the time stack data were collected every day at 08.00GMT, 10.00GMT, 12.00GMT. On June, the collection frequency was increased to become hourly schedule, running from 06.00GMT until 14.00GMT. Video images which are taken later than 14.00GMT (1600 local time) were usually severely contaminated by sun glare.

As ground truth bathymetry, a field survey was performed between 4 June and 7 June 2013 comprising sub-tidal bathymetric survey using echo-sounder mounted on a jet ski combined with intertidal mapping which involved walking survey equipped by GPS recording devices (e.g. see Van Son *et al.*, 2009 for personal water craft-based bathymetric survey). The survey spatial sampling was made fine enough, with various spacing, starting from 10 m close to the shore (e.g. around the most rhythmic bar), 25 m at the surf and shoaling zone, and 50 m at the offshore.

To ensure smooth bathymetry and avoiding gaps, the running average will be initiated using bathymetry surveyed from August 2011. This initial bathymetry is set to have a relatively high standard deviation error of 5 meters (e.g. low confidence as they are old bathymetry data). Surveyed bathymetry will be compared with cBathy running average result from 7 June 2013 1400GMT. The results are presented in Figure 5-9. The cBathy estimate (left) shows a very good agreement with surveyed data (middle). The outer sand bar at approximately x = 400 m is produced very well. Moreover, features near the shoreline produced by the cBathy resemble the surveyed data adequately. To evaluate the agreement in more detail, cross shore profile comparison is presented in Figure 5-10. Bar crests and troughs are predicted very well by cBathy (left panel). Moreover, the alongshore bar crest transect (right panel) also shows high significant agreement between cBathy estimate and surveyed bathymetry. However, close to the shoreline, depths are highly overestimated, which is due to the shallow depths, therefore the cameras capture mainly the swash motions instead of wave motions. Moreover, analysis tiles near the shore may include 'dry pixels' and therefore provide faulty estimates of water depths (Holman *et al.*, 2013). This will be discussed and mitigated in Section 5.4.4 and Section 5.5. The global root mean square error of the cBathy estimate is 0.88 m.

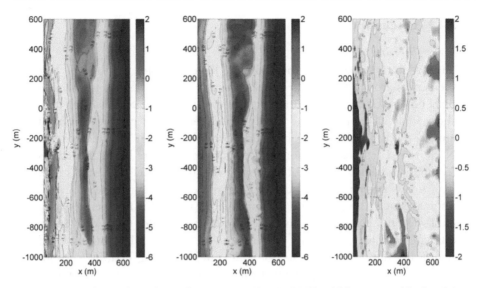

Figure 5-9: Left: cBathy estimate for 7 June 2013 1400GMT, middle: surveyed bathy, right: difference plot (cBathy minus surveyed).

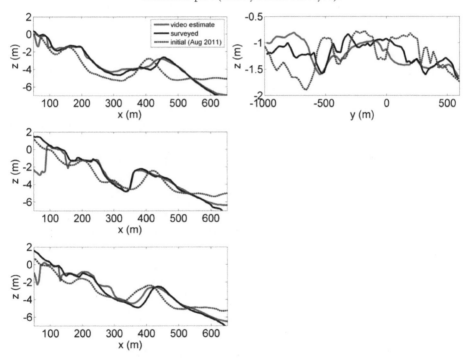

Figure 5-10: Left: top to down-cross shore profiles of bathymetry at y = 400, -200, and -600 meter. Right: alongshore transect over the bar crest. Red: cBathy, black: data, blue: initial.

91

5.4.4 Discussion

Near-shoreline problem
Shallow water depths near the shoreline have degraded cBathy skill to estimate the depths around this area. Here, an analysis is carried out to test whether using pixel time stack data only during high tide will improve cBathy estimates. Results show, that when pixel time stacks data used in the cBathy computation are limited to data only during high tide condition (when the tide is greater than NAP 0 m), the overestimation of depths near the shore line is indeed decreased. In Figure 5-11 profile plot of cBathy results using pixel time stacks only during high tide is presented (note the cross shore extent now limited to x = 300 m). Estimates near the shore line are improved. Around the northern edge (y ~ -200 until -800 m), using pixel time stacks only during high tide clearly improve the estimates closer to the data (green lines in Figure 5-11), however still show significant differences. In section 5.3, integration of cBathy (sub-tidal) with intertidal bathymetry obtained from shoreline detection technique will be discussed to mitigate this issue.

Parameter for the process error
In the Step 3 of the cBathy process, a process error variable, Q, is introduced in addition to the error variance of the running average (Equation 5-17). This process error is modelled in such a way that the error will build up during the un-modelled period, and is a function of wave height and spatial location as given in the Equation 5-19. The process error curve is centred on the location where the most dynamic bed changes are expected. In this case, x = 220 m is chosen (approximately the crest of the inner bar, see Figure 5-11, dark lines). The parameter σ_x gives the spread of the curve and the C_Q is a constant. For Egmond application presented previously, best guest values for σ_x and C_Q is 75 m and 0.04/day respectively. In cBathy paper, data set from Duck, NC are used to calibrate for C_Q as there are no literatures provide sufficient information on this, and the data set is the best data set available to do so (Holman *et al.*, 2013). Here, an investigation is performed to evaluate optimum parameter σ_x and C_Q that gives lowest error between estimate and surveyed bathymetry. The parameter σ_x was made varying, from 5 to 125 meters, while C_Q was ranged from 0.01 to 0.07. The same running procedure with one presented in Section 5.4.3 was performed for each combination of σ_x and C_Q to obtain cBathy estimate for 7 June 2013. The results are presented in Table 5-2. Using different C_Q only affect the rms error slightly. For σ_x = 50, 75, 100, and 125 meter, higher C_Q gives higher rms error with maximum difference (between C_Q = 0.01 and C_Q = 0.07) is 10 cm. On the other hand, for σ_x = 25 m and lower, different C_Q results in approximately constant rms error of 0.75 m.

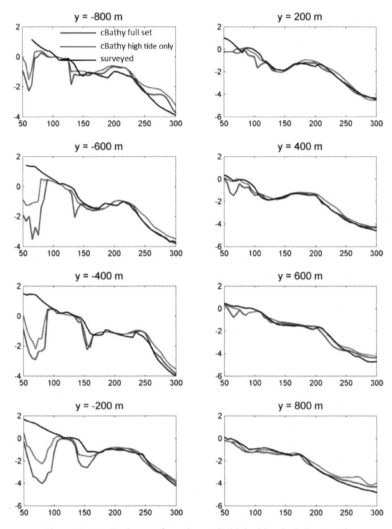

Figure 5-11: cBathy results using only high tide pixel time stacks

Table 5-2: cBathy rms error for different σ_x and C_Q

	rms error (m)						
	$\sigma_x = 5$	$\sigma_x = 10$	$\sigma_x = 25$	$\sigma_x = 50$	$\sigma_x = 75$	$\sigma_x = 100$	$\sigma_x = 125$
$C_Q = 0.01$	0.75	0.75	0.75	0.76	0.81	0.85	0.87
$C_Q = 0.02$	0.75	0.75	0.75	0.76	0.84	0.89	0.91
$C_Q = 0.03$	0.75	0.75	0.75	0.77	0.86	0.91	0.93
$C_Q = 0.04$	0.75	0.75	0.75	0.78	0.88	0.92	0.95
$C_Q = 0.05$	0.75	0.75	0.75	0.78	0.89	0.94	0.96
$C_Q = 0.06$	0.75	0.75	0.75	0.79	0.9	0.95	0.97
$C_Q = 0.07$	0.75	0.75	0.75	0.79	0.91	0.95	0.97

5.5 Integration of sub tidal bathymetry from cBathy with intertidal bathymetry from shoreline detection method

5.5.1 Approach

In order to obtain a smooth and accurate bathymetry estimate especially near the shoreline, cBathy is combined with intertidal bathymetry estimates mapped based on detected shorelines from time exposure images. The method, Automatic Shoreline Mapper- ASM, described in Section 5.2.3, is used to obtain the intertidal bathymetry, generated from shorelines collected from the last three days of each time stamp of the pixel stack data (automatic detection and without human supervision).

In order to avoid overestimation of depths with cBathy near the shore line, the pixel time stacks will be treated in such that only wave motion like signal given by the pixel intensity will be analysed by the algorithm. To this end, an elevation contour, acts as a separation line, is introduced towards the full set of the pixel time stacks based on actual tidal elevation plus a minimum depth criterion. The elevation is defined as:

$$z_C = z_{tide} - h_{min} \qquad\qquad 5\text{-}20$$

, where h_{min} is taken 0.5 m, assuming typical mean wave period T = 5.5 seconds and minimum wave length resolvable by the video sampling resolution at Egmond is approximately 12 m. To avoid having different spatial domain every time stamp due to the negation of data near the shoreline, constant value of one are assigned to the area above the separation line as flags. By doing this, the cBathy algorithm will easily detect signal from this area as "bad signals" instead of coherent swash signals; therefore will return NaN as estimates. This rectified pixel time stack then will be used in cBathy algorithm in Step 1 and Step 2 (Section 5.4.1). The estimates from Step 2 will then be merged with intertidal bathymetry obtained from the ASM. And finally, this merged estimate will be smoothed and weighted in Step 3 of cBathy, which is the running averaged (Kalman filter) phase. However, to objectively weight and combine these two bathy estimates, a corresponding error (P in Equation 5-16 until 5-18) of ASM estimates is needed. Here, a constant value of 1 meter is used. As described in Section 5.4.3, a surveyed bathymetry from August 2011 is used as initial running average bathymetry with corresponding error of 5 meters. This implies that the separation line previously described will also be based on this old bathymetry at initial stage of the running average process. The results however converged appropriately until the last pixel stack data instead of showing bathymetric gaps in the upper face of the intertidal part. This suggests the initial separation line, even obtained from old bathy data, provides decent estimates. The procedure is summarized in a workflow presented in Figure 5-12.

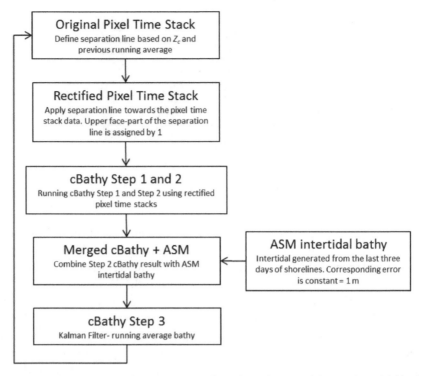

Figure 5-12: Flow diagram for integration of cBathy estimates with ASM intertidal bathy

5.5.2 Results and discussions

The integration technique was tested using pixel time stacks collected from April until June 2013, similar to the ones described in Chapter 5.4.3. Including intertidal bathymetry estimates from ASM significantly improve the bathymetry estimate near the shore line. In Figure 5-13, bathymetry plot of the default cBathy and the ones combined with intertidal bathymetry are presented. Distinct improvement is clearly seen in which combined cBathy and ASM (Figure 5-13b) smooth the estimates of the default cBathy (Figure 5-13a) near the shore with appropriate values resembling the surveyed data. A bathymetric feature that appears like a small headland at y ~ 700 m predicted well as it was not clearly present in the default cBathy. The improvement becomes clearer in the cross shore profile plots. In Figure 5-14, cross shore profiles for different alongshore locations are presented. Improvement is highest for the northern part at y = - 200, -400, -600, and -800 m (from the red line to become the blue line). For transects at y = 200, 400, 600, and 800 meter, the improvement is less, however show the best agreement with surveyed data. This is due to the relatively deeper contour of this area compared with the area on the north.

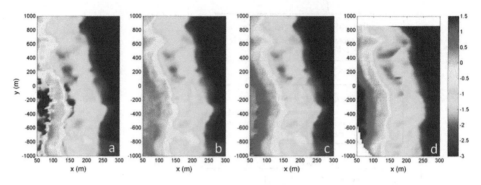

Figure 5-13: Bathymetry comparison, (a) cBathy, (b) cBathy + ASM, (c) cBathy + IBM, (d) surveyed bathy.

Adding intertidal estimates from ASM to the cBathy clearly improve overall estimate of the nearshore bathymetry. The rms error for the sub domain like shown in Figure 5-13 is 1.17 m for "cBathy only" estimates (a) while merging with intertidal bathymetry from ASM significantly decreases the rms error to 0.38 m (b). While the bathymetric profiles improved a lot, the upper part of the intertidal bathymetry still show overestimation of depths. A further test is performed by using intertidal bathymetry generated from user-supervised waterlines (IBM, see Section 5.2.3). In Figure 5-15, intertidal bathymetry generated using ASM (un-supervised waterlines production, red lines) is compared with intertidal bathymetry generated using IBM (user-supervised waterlines production, blue lines). The period from which the water lines were generated is from 3 June until 7 June 2013 (e.g. manual supervision in generating shorelines every 3 days). The result shows that intertidal bathymetry from IBM predict the upper face part better than ASM. The improvement is best for transects at y = -200 until y = - 600 meters. Applying this intertidal bathymetry into the cBathy integration clearly improves the estimate for the upper face of the intertidal area (see Figure 5-13c and Figure 5-14 green lines). The rms error for this sub domain slightly decreases from 0.38 m (ASM based) to become 0.34 m (IBM based).

Although the estimate from combined cBathy-IBM shows better agreement with surveyed data, IBM based intertidal bathymetry requires man hours for supervision in order to approve/reject detected waterlines, which is costly from automatic-operationalization point of view. For instance, to collect waterlines from three days set of video images (the last three days waterlines-based intertidal bathymetry) will require approximately 3 to 4 hours of work for an experienced user. This is less favourable from operational point of view. On the other hand, ASM based intertidal bathymetry shows great potential of application in an autonomous fashion, however overestimates depth at the upper face of the intertidal bathymetry due to the need of manual shoreline data for restarting (Uunk *et al.*, 2010). Improving the shoreline detection technique in such restarting is less required will add to not only the skill of the integration technique described here, but also the potential of the approach to be implemented for operational application.

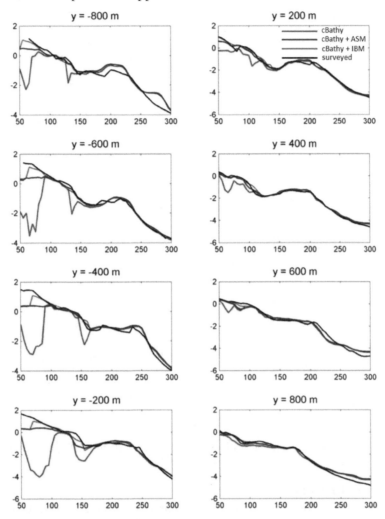

Figure 5-14: Cross shore profile of bathymetry comparison for different combination of bathy sources

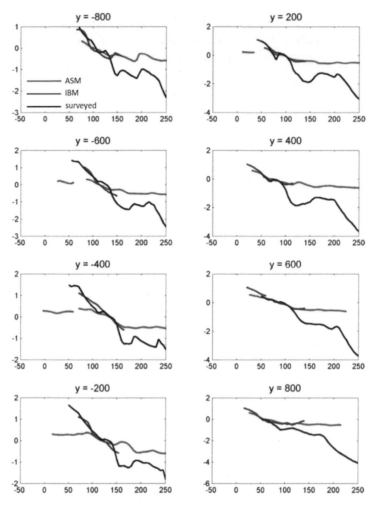

Figure 5-15: Profiles of bathymetry obtained from ASM (red) and IBM (blue) using waterlines from 3 June until 7 June 2013

5.6 Conclusions

Nearshore scale bathymetry can be accurately obtained from video camera. Beach Wizard-dissipation maps and cBathy both show a good skill in predicting nearshore bathymetry. Using wave dissipations maps as sources, Beach Wizard can nowcast bathymetry including important rip channel features as seen in the surveyed bathymetry. With cBathy, bar crest location and trough are very well predicted, while Beach Wizard performs less on these areas. More potential future application is shown by cBathy. Unlike Beach Wizard which needs a set of images over a certain period of time to produce a single bathymetry, cBathy can deliver a bathymetry from a single set of pixel time stacks, thus can provide bathymetry estimates (almost) every hour. Moreover, unlike Beach Wizard dissipation maps, cBathy does not need an image selection procedure since the algorithm is designed to be intelligent enough to handle bad properties of images.

An appropriate set up of cBathy pixel collection and parameter settings have been set up for Egmond aan Zee site, as well as a strategy to handle the poor estimates returned by cBathy around the dry pixel areas. Integrating cBathy estimates with intertidal bathymetry obtained from shoreline detection technique significantly improve the bathymetry estimate near the shore line, with the global rms error decreasing from 1.05 m to 0.36 m. While the intertidal bathymetry obtained from a set of user-supervised water lines (IBM) gives better results compared with the one from ASM, implementing IBM in operational mode is costly because it requires man hours for supervision. Therefore the analysis in the thesis will be continued using the intertidal bathymetry from ASM. Further on in this thesis, cBathy estimates will be used as nearshore bathymetry boundary to be applied in the numerical modelling in order to forecast rip currents (Chapter 6).

6 Predicting rip currents: Combination of CoSMoS and bathymetry from video

6.1 Introduction

In this Chapter, the applicability of bathymetry obtained from video technique to be used as input for hydraulic modelling will be presented, followed by an example of the operational forecasting type of simulation on predicting rip currents at Egmond aan Zee.

Firstly, in Section 6.2, the applicability of bathymetry from video for nearshore current predictions using numerical modelling is analysed by comparing model results using surveyed bathymetry and model results using video bathymetry as input. Next, in Section 6.3, a test case will be presented where bathymetry from cBathy will be used in the CoSMoS model system in a forecast mode, to simulate rip currents. Field data on the observation of rip current events at Egmond aan Zee are available for the period of August 2011 (Chapter 4). However, during that period, pixel time stack collection has not been set up yet, therefore cBathy estimates cannot be produced in this period. cBathy estimates are available from March 2013 onwards, therefore, as a test case, the summer 2013 period will be taken. Reports on rip related incidents during the month of August 2013 are available through the Twitter page of the lifeguards at Egmond aan Zee. This incident information will be used to verify qualitatively the forecast results.

6.2 Applicability of cBathy bathymetry on the prediction of nearshore currents

The aim of this section is to analyse the applicability of cBathy bathymetry to be used in numerical modelling in order to predict nearshore currents. First, model results and field data are compared to verify that the main processes are simulated. Second, the applicability of video bathy from cBathy in predicting nearshore currents using numerical model will be analysed by comparing it with model results using ground truth bathymetry as input.

6.2.1 Comparison with field data

To verify that the main processes are simulated with the model, Eulerian measurements of currents and surface elevation will be compared with simulation results. This point measurement utilizes sensors, which were mounted on a frame at a fixed location, at a water depth of 1.6 meters from MSL (Figure 6-1). Data were acquired at 16 Hz sampling with a continuous recording from 4 June until 7 June 2013. The data obtained from the sensor were filtered in such only measurements that pass certain quality controls will be used in the analysis. Wave conditions observed at IJmuiden wave buoy for the period of simulation were mild, with significant wave height varying from 0.6 to 1.4 meters, and peak period varying between 4 and 9 seconds. During this period, the wave direction was consistently north and northeast (see Figure 6-2). This condition results in a dominant alongshore current inside the surfzone towards the south.

CoSMoS is used to provide boundary conditions for the Egmond aan Zee model. The model chain is presented in Figure 6-3. The Egmond model (the last panel on the right) acquires boundary conditions from Kuststrook model, which is a model nested into a bigger model, The North Sea model (see Chapter 3). The spatial domain for Egmond aan Zee model covers approximately 1.5 km alongshore and 700 m cross shore. The hydrodynamic time simulation starts on 3 June 2013 until 8 June 2013. Field data are available from 4 June, therefore providing ample spin-up time for the model. A term 'ground truth model' and 'video bathy model' will be used herein to refer to simulation using surveyed bathymetry and simulation using cBathy bathymetry, respectively.

Wave height, water level, and velocity components are collected from both models (ground truth and video bathy model), and compared with data. The time series results are presented in Figure 6-4. The ground truth model and video bathy model results are in good agreement with the data. Water levels can be predicted well over the simulation period. Model-data agreement for wave height is slightly less, especially after the low tide (towards the rising tide). Root mean squared error and bias for the wave height is 0.09 m and 0.08 m, respectively (both for ground truth and video bathy model). With an average wave height of 0.4 m during the simulation period, the rms error suggests a normalized error of 22.5%.

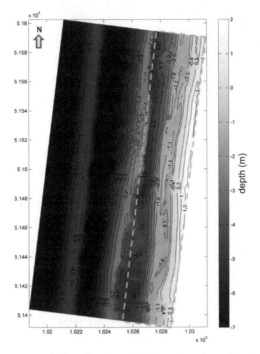

Figure 6-1: The measurement location for velocity and surface elevation (red point), and nearshore sub domain for analysis (dashed line).

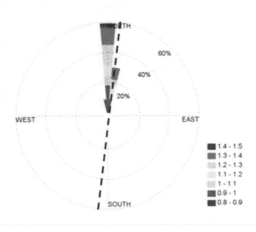

Figure 6-2: Wave rose from IJmuiden wave buoy during the field work. The black dashed line is shoreline orientation (see Figure 4-1 for the buoy location).

Figure 6-3: Model chain within CoSMoS, showing model domains starting from the entire globe, the Dutch Continental Shelf, the Dutch Coast, and finally Egmond aan Zee.

The modelled and observed cross shore component velocity u shows a consistent small magnitude over the simulation period, and models show a good agreement with data (bottom panel in Figure 6-4). In contrast, the modelled alongshore velocity component, v, shows a consistent alongshore current towards the south. The observed alongshore current magnitude varies between 0 to ~ -0.4 m/s (negative, flowing to the south). This dominant process was simulated well by both ground truth and the video bathy model. The rms error is 0.1 m/s and 0.13 m/s, with a bias of 0.01 m/s and 0.02 m/s, for ground truth model and video bathy model respectively. With an average alongshore magnitude during the simulation period of 0.34 m/s, these rms errors suggest 29% and 38% normalized error for the ground truth and the video bathy model respectively.

6.2.2 Nearshore currents, model-model comparison

The performance of the cBathy bathymetry will be evaluated by comparing flow fields generated by the model obtained from simulations using ground truth bathymetry. Results show that nearshore current flow field simulated using cBathy agree very well with simulations using ground truth bathymetry. In Figure 6-5, rms error and bias of the velocity magnitude between the two simulations are presented and are grouped per bin depth. Over the whole model domain, agreements of the simulated currents are very good. The best agreement is found in the areas with water depth greater than 4 meters, where the rms errors are less than 0.03 m/s, with bias between -0.01 and 0.01 m/s. In shallower areas, rms errors are higher, and the highest rms error is found in the shallow bin depth. Temporal variability of the errors can be analysed by looking at the effect of varying water level due to the tide on the errors of the model results. In Figure 6-6 time varying rms errors is presented. The rms errors are averaged over the sub-domain (small box) displayed in Figure 6-1. In accordance with previous bin depth analysis, the rms error is found to tend to increase during lower water level and decrease as the tide is rising, but overall, showing a good agreement between the two model results.

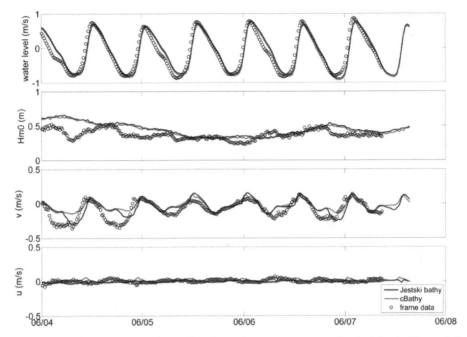

Figure 6-4: Model data comparison taken from the measurement point (red dot in Figure 6-1), blue: ground truth model, red: video bathy model results, black circle: data.

On the Dutch coast, tidal asymmetry is distinct, in which rising tide occurs faster than falling tide. During floods, the tidal current flows to the north, and flows back to the south during ebb. The magnitude of the tidal current is greater during flood compared with the magnitude during ebb. These can be seen from the simulation results, and cBathy simulation shows a good agreement with the ground truth. In Figure 6-7 and Figure 6-8, nearshore currents simulated by the models using different bathymetry sources are presented, for the flood and ebb period, respectively. The cBathy simulations (right panels in both figures) exhibit similar nearshore flow fields with ground truth simulation (left panels). Two different flow regimes due to the tidal current and wave driven current which flowing in opposite directions can be clearly seen from both simulation results with different bathymetry sources (Figure 6-7).

The difference plot between the two model results is presented in Figure 6-9. During the ebb, differences (colour map) are slightly stronger compare to the one during the flood. In the intertidal area, differences are found to be big for the flood period. This is due to the overestimation of intertidal bathymetry predicted by the Auto Shoreline Mapper discussed in Chapter 5.

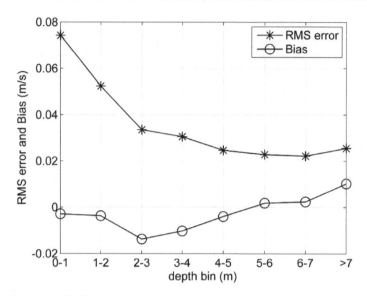

Figure 6-5: RMS error and bias for the velocity magnitude per bin depth

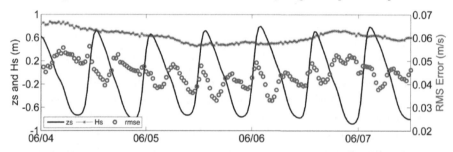

Figure 6-6: RMS error (red circle) of velocity magnitude between ground truth model and cBathy model results, averaged over the sub domain. The black line is tidal water level and the blue markers are offshore wave height.

Average over the sub domain, the rms error of velocity magnitude between the two model results is 0.05 m/s and 0.04 m/s for ebb and flood respectively. One can assume a typical (alongshore) nearshore currents for Egmond aan Zee of ~0.5 m/s, therefore the relative rms errors are 10% and 8%, showing a very good agreement between ground truth simulation and video based one in simulating nearshore currents.

Results presented in this section verify that using the bathymetry from cBathy, the hydraulic model can predict nearshore currents as adequately as when surveyed bathymetry is used. Nearshore currents predicted by cBathy simulation agree well with ones using ground truth bathymetry. However, ground truth data available to be used for such verification are limited.

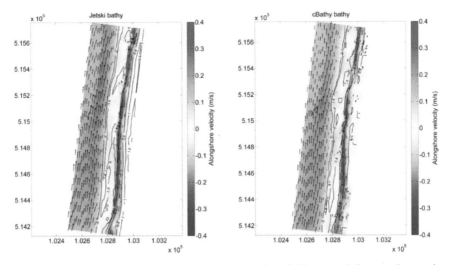

Figure 6-7: Alongshore velocity (colour-map) and flow field (arrows) for nearshore sub domain, averaged over flood period, left: ground truth model, right: video bathy model.

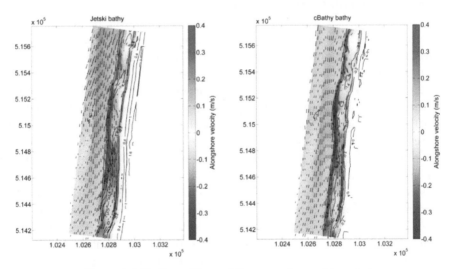

Figure 6-8: Similar to Figure 6-7, averaged over ebb period

In addition, environmental conditions during which this comparison is tested exhibits a more or less uniform flow field in alongshore direction. In the future, validation using bathymetry condition comprises rip channel dynamics and varying incoming wave angles would provide us with a better impression on applicability of video bathymetry to be used in nearshore model, especially in predicting rip currents.

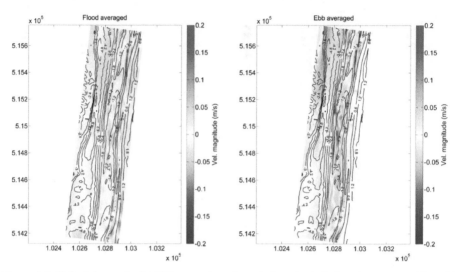

Figure 6-9: Velocity magnitude difference over the sub domain, left: flood averaged, right: ebb averaged.

6.3 A test case: summer 2013

In this section, the rip current forecasting system will be tested, where we combined CoSMoS model system in forecast mode and cBathy bathymetry to provide the operational type of simulation. Data regarding rip current events for Egmond aan Zee are not available for this period. However, there is some qualitative information related to rip current incidents available through the Twitter page of Egmond Life guard (https://twitter.com/RB_Egmond, Figure 6-10). In the timeline of the web page, the life guard frequently post tweets regarding safety and actual information related to the beach, the community around it, and recreational issues. Tweets regarding rip currents, for instance, when the weather is not favourable and visitors are not allowed to swim are also posted. Occasionally, rescues to prevent drowning are also posted. In addition, information about warnings which are given by flag system can also be obtained visually from the Argus images, since the office of the lifeguards where the flag that is usually flown is well covered by the camera. The combinations of the two aforementioned sources of information are summarized in Table 6-1. The wave heights considered in the table are averages over a particular day between 06:00 and 18:00 GMT, recorded at the IJmuiden wave buoy. Starting from 29th of July, a red-flag *tweet* was posted, which means swimming is not allowed due to unfavourable wave conditions. Afterwards, on 4th, 10th, and 14th of August there were incidents (followed by rescues) reported and all happened when the warning-flag is *yellow* (yellow means swimming is dangerous and floating swimming gears are not allowed). Moreover, during relatively calm weather, *orange* warnings were issued, which means bathing or swimming is allowed but floating swimming gears are not. This information will be used to qualitatively test the proposed model system in predicting rip currents.

According to the lifeguards at Egmond aan Zee (pers. Comm.), rip channelled bathymetry usually start to develop around July or August, and will stabilize over the summer. This information supports cBathy estimates. In Figure 6-11, bathymetric evolution from June (when surveyed bathymetry was available, Chapter 5 and 6.2), until mid of August are presented, for several representative dates. We can clearly see that rip channel like patterns were not clearly observed during June and became clearer on July and August (dark arrows). The inner bar, located around x = 200 m can be clearly seen interrupted by channels for the month of July and August. Since there are no surveyed bathymetry data available to verify this, the cBathy estimates are visually compared with time exposure images. In Figure 6-12, comparison of cBathy estimates and time exposure images for 3 different dates are presented. The rip channel features estimated by cBathy are also observed visually from the time exposure images. Features which are combination of *brighter-darker* colour corresponds to *shallower-deeper* depth is clear from the time exposure images and cBathy estimates present similar patterns. This qualitative comparison result suggests cBathy estimate exhibit the skill in predicting actual bathymetric features for the month of July and August.

Figure 6-10: A print screen of Egmond's life guard Twitter web page.

Table 6-1: Summary of rip current warnings.

Date	Incident	Flag	Mean H (m)	Max H (m)
29 July	-	Red	1.15	1.69
1 Aug	-	Orange	0.5	0.75
2 Aug	-	Orange	0.31	0.38
3 Aug	-	Red	1.43	1.83
4 Aug	Yes	Yellow	0.53	0.69
5 Aug	-	Orange	0.35	0.39
6 Aug	-	Yellow	0.8	1.15
9 Aug	-	Yellow	0.85	0.91
10 Aug	Yes	Yellow	1.29	1.59
11 Aug	-	Red	1.2	1.5
12 Aug	-	Yellow	1	1.24
13 Aug	-	Red	1.47	1.68
14 Aug	Yes	Yellow	1.21	1.32

The CoSMoS system was set up to generate boundaries for Egmond aan Zee model for simulation time starting from 1 August until 14 August 2015, covering the period of incident reports from the life guards' Twitter web page. To provide operational type of simulation, meteorological data used in the CoSMoS simulation are 48 hour forecast data. Moreover, the bathymetry boundary to be used by the Egmond local model will be the daily cBathy estimates. XBeach is used for the local model to simulate waves and hydrodynamic, which was set up similarly with one presented in Chapter 4. Time varying bathymetry boundary, which is obtained from cBathy estimate, is applied using keyword *setbathy*. This approach is for experimental purposes, as in the operational mode, cBathy will provide bathymetry to the system in online mode.

In Figure 6-13, forecast offshore directed velocity is presented, together with the warnings provided by the lifeguards. During the period of analysis, rip currents were predicted almost every day and show a tidal modulation. This prediction is supported by the warnings given by the lifeguards, which during this period, orange, yellow, and red code of warnings were issued. However, there are days (7 and 8 of August) where rip currents were predicted, but there was no warning. For 7th of August, there were much less people seen (from the Argus cameras) on the beach, because the weather during that day was rainy.

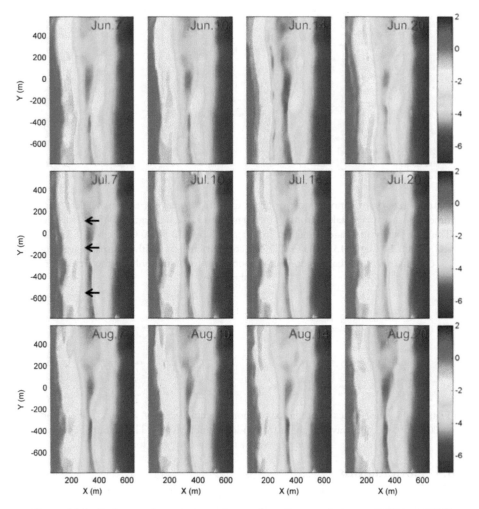

Figure 6-11: cBathy running average estimates from June until August 2013 in ARGUS coordinates for several representative dates.

The results of the forecast during which rescues were reported, are shown in Figure 6-14 as phase averaged plot. The panels b), c), and d) clearly predicts rip current events in 4, 10, and 14 of August 2013 respectively. The reddish colour code indicates offshore directed velocity, and the vector plot means the flow field. The strength of the rip is much less during 4 of August since the average wave height was also lower compared with 10 and 14 of August. For comparison, the forecast result for 7 of June 2013 (a) is also presented, during which offshore directed flows were not predicted by the model system. Supported by the previous analysis presented in Chapter 4, the offshore directed flows were predicted during the low tide, while during the high tide, the rip currents were absent. The results suggest that rip currents at Egmond aan Zee can be forecast, which can be used for warning purposes.

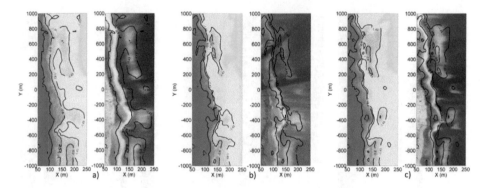

Figure 6-12: cBathy estimates versus time exposure images. For each panel: left: cBathy, right: time exposure image, taken from a). 02 August 2013, b). 10 August 2013, c). 14 August 2013. The time exposure images are taken from a). 31 July 2013 08.00 GMT, b). 10 Aug 2013 09.00 GMT, c). 14 Aug 2013 10.00 GMT. The contour lines of cBathy estimates are overlaid for each figure. Note that the domain is zoomed in focussing only on the area close to the shoreline.

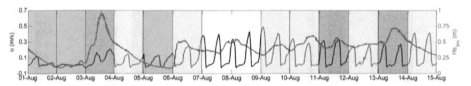

Figure 6-13: Time series of predicted offshore directed velocity, u (black), taken from (x, y) ~ (200,-150) m, and wave height at the model boundary (red). The colour shadings indicate warnings given by the lifeguards (see text).

6.4 Conclusions

The first part of this chapter discusses applicability of using cBathy bathymetry in numerical models to predict nearshore currents. The results suggest cBathy bathymetry shows great potential to be used as a bathymetry boundary in numerical modelling. Nearshore currents predicted by the ground truth model agree very well with ones predicted by the cBathy bathymetry model. The rms error of velocity magnitude between the two model results varies in time, where during ebb rms error is slightly higher compared with the flood period. Over the whole simulation period and the nearshore subdomain considered, rms error is 0.044 m/s, indicates great potential of cBathy to be applied in numerical model for bathymetry boundary as surveyed one is used.

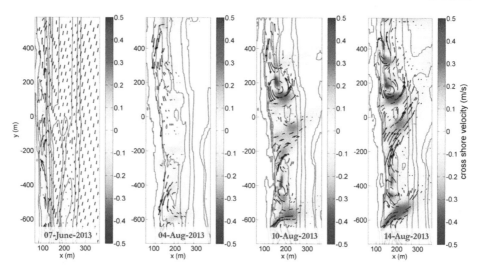

Figure 6-14: Rip current forecast for 4 (b), 10 (c) and 14 (c) of August 2013 correspond to the dates when rip incidents were reported through the life guard's web page. For comparison, result from 7 June 2013 is also presented (a) when rip currents were not predicted and ground truth bathymetry was available (Chapter 6.2). The colour code is cross shore velocity (positive offshore) averaged during low tide, the arrows are flow field, and the grey lines are bathymetry contour from cBathy daily estimates.

The second part of this chapter presents a test of rip current forecasting. A CoSMoS simulation for Egmond aan Zee beach was initiated for the first weeks of August 2013, during which some rip current incidents were reported by the life guard. Applying CoSMoS in forecast mode, rip currents were predicted for the period of the incident events. Dangerous rips were predicted to occur during the low tide, which is supported by the previous analysis presented in Chapter 4. This confirms the potential application of the proposed CoSMoS-cBathy system in providing forecasts for rip currents at Egmond aan Zee.

7 Summary and outlook

7.1 Summary

Rip currents are among the most dangerous coastal hazards for the bathing public, and contribute to the highest portion of beach rescues all over the world. In order to help life guards in planning and preparing rescue resources so as to minimize casualties, information about where and when rip currents may occur is needed. This can be provided by a predictive tool which combines meteorological forecasts, hydrodynamic models and remote-sensed observations. However, to implement this approach for the nearshore at a beach resort, up-to-date and high resolution bathymetry data are needed since the time scale of nearshore morphological change can be in the order of days to weeks, depending on the environmental (waves, hydrodynamic) conditions. Therefore, having accurate bathymetry data based on conventional bathymetry surveys on this time scale would require a large logistical effort, and would be very costly.

The objective of the current research is to develop and test a methodology with which forecasts of rip currents can be provided for swimmer safety purposes at Egmond aan Zee. An operational model system, CoSMoS (Coastal Storm Modelling System, Van Ormondt *et al.*, 2012), is used as the main task manager to control the operation of numerical models. A validation of CoSMoS has been carried out to evaluate the performance of the model system in providing waves and water level boundary conditions. To evaluate the model's ability in simulating rip currents at Egmond aan Zee, numerical experiments simulating rip currents were set up and the results were validated using data from field experiments. Furthermore, the numerical experiments were extended, and a longer period was simulated in order to gain more knowledge on rip current characteristics at Egmond aan Zee and the safety implications. To meet the need of continuous and up to date bathymetry, a technology using video images to predict nearshore bathymetry on a daily basis, cBathy (Holman *et al.*, 2013) was utilized.

Three research questions were formulated in order to test the new proposed approach:

1. Can we predict the occurrence, duration, and magnitude of the rip currents at Egmond using a processed-based model? What is the added-value for swimmer safety warning systems?
2. Can we obtain nearshore scale of bathymetries through video technique for Egmond aan Zee in an operational mode?
3. Can we apply nearshore bathymetry from video to predict nearshore currents and to forecast rip currents?

In the following, each of these research questions will be summarized based on the results from the thesis, and outlooks towards further improvements will be elaborated.

7.1.1 Can we predict the occurrence, duration, and the magnitude of the rip currents at Egmond using process-based model? What is the added-value for swimmer safety warning systems?

To answer this question, XBeach model was utilized to investigate rip current initiation and duration. The model was validated against Lagrangian velocity measurements obtained using GPS floating drifters from a field campaign, where rip current events were depicted very well from the data set. Validation results show that the model exhibits good agreement with data, where several flow pattern types of drifters flowing through the rip channels were produced well by the model. From the numerical experiments, it is found that the rip currents at Egmond aan Zee are driven by wave action and the initiation and duration are strongly controlled by water level. For the period of analysis, the rip currents start to initiate approximately 5 hours before low tide, reach their peak during (peak) low tide, start to decay as the tide is rising, and finally become inactive 3 hours after low tide. Their initiation corresponds to the ratio of offshore wave height to water depth on the bar of ~0.55. Rips may also occur during the high tide (when the 0.55 ratio is fulfilled), which requires a relatively high wave height.

Furthermore, from the numerical experiments, at Egmond drifters transported offshore by the rip are unlikely to circulate back to the shore (only 14% do so) contrary to observations at other beaches. The drifters rather stay outside (e.g. contour depth of more than 2 meters), where they are further advected alongshore by tidal currents. This has implications for safety: the common guideline for swimmers once caught in a rip current is to swim parallel to the beach. For the Dutch Coast, except during slack water, the tidal current will be dominant outside the surf zone and swimming parallel to the shore will only be the correct strategy when the swimmers swim into the right direction (with the current). Otherwise, the impact can be as dangerous as swimming against the rip. For Egmond aan Zee, a strategy of 'do nothing' and signalling for help is proposed, since the rip currents are unlikely to transport the swimmers back to the shore whilst strong tidal current is likely to be present. In addition, the 'do nothing' strategy can also be combined with swimming back to the shore, because after 'letting go' for e.g. 5 to 10 minutes, one is most probably already outside the offshore flow path of the rip, especially when the rip's orientation is oblique, therefore swimming back onshore can be an efficient strategy.

From the numerical experiments, it was also found that within 5 minutes, an object (drifter) can be transported as far as 60 meters offshore from the rip channel, even 4 hours before the low tide. As the tide is approaching the low tide, this number increases to ~90 m during the (peak) low tide. These findings define the threat of the rip at Egmond aan Zee.

7.1.2 Can we obtain nearshore bathymetries through video technique for Egmond aan Zee in an operational mode?

Nearshore bathymetry can be accurately obtained from video imagery. In Chapter 5, two techniques with which nearshore bathymetry can be obtained using video data were tested. The first technique, Beach Wizard, shows reasonable skill in predicting bathymetry using wave dissipation maps derived from time exposure images. The technique was able to predict the alongshore bar-channel pattern, which is important for rip channel identification. However, the skill of the technique is less in predicting bar-trough in cross shore orientation. Moreover, availability of good dissipation maps from video may be scarce in time, since not only good video images are needed, but also good signature of wave dissipation, which is strongly dependent on actual wave heights and water level conditions. On the other hand, cBathy, which is based on estimating wave celerity, performs very well both in alongshore and cross shore orientation, and is designed to provide daily estimates of bathymetry. The bathymetric features predicted by cBathy are in very good agreement with data, including the bar crest and trough location and elevation. In the very shallow area near the shorelines, cBathy performs poorly; however, this can be mitigated by integrating cBathy estimates with intertidal bathymetry obtained from shoreline detection techniques. The integration significantly improves the bathymetry estimates at the very shallow water. Therefore, from an operational point of view, cBathy shows a great potential to be applied.

7.1.3 Can we apply nearshore bathymetry from video to predict nearshore currents and to forecast rip currents?

Bathymetry obtained from video technique, cBathy, can be applied in a numerical model, just as a surveyed one is used. Results show that nearshore currents simulated using video bathy agree very well with those using surveyed/ground truth bathymetry. Coupling the video bathymetry estimates with CoSMoS to obtain a nearshore forecast system on rip currents has also been tested. In forecast mode simulation, dangerous rips were predicted which were verified qualitatively using rip incident report posted by the lifeguard through Twitter page. This can be achieved through the capability of cBathy to predict rip channel features. Applying the video bathymetry to operationally forecast rip currents therefore is feasible.

7.2 Outlook

7.2.1 Application of the system and how the information can be useful

For operational application, the CoSMoS system is applied to provide forecasts of rip currents and to present the graphical results of the model predictions. The types of information that can be presented are the spatial distribution of the currents over the coastal stretch like the ones presented in Figure 6-14. In addition, a more aggregated form of the information can also be displayed, as shown in Figure 7-1 (Van Ormondt *et al.*, 2012), where time is on the x-axis and location alongshore on the y-axis. This type of information will be useful to the life guards and the public. The location and time where a rip current might occur is shown in a colour-shaded code indicating its strength. The bottom panel shows the tidal elevation as a function of time, where the colour shades indicate the maximum strength in the alongshore. This makes rip current locations, timing, and the strength become very predictable. With this kind of information, the local life guard organization will be assisted in allocating and planning their resources and providing warnings to the public.

CoSMoS is a generic system and able to accommodate different types of model suites. In the initial state of the CoSMoS development, a coupled tide-phase averaged wave model of Delft3D was applied for the local high resolution model. In this thesis, XBeach is used as the local model since the spatial distribution of flow field in the vicinity of the rip channels is better predicted when wave group forcing is included (Chapter 4). However, this requires a costly computational time, which may hamper its applicability for operational forecasting application. With recent improvements in computing speed of XBeach and expected further increase of computer power, this problem is likely to be resolved in the near future.

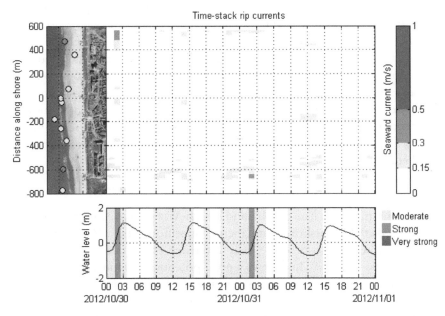

Figure 7-1: Alongshore time stack of offshore directed velocity (Van Ormondt *et al.*, 2012, http://muienradar.nl)

7.2.2 Future research topics

Rip currents field data and swimming safety

During this study, Lagrangian data obtained using floating GPS drifter were used for validation materials towards the XBeach model. While some of the data clearly show offshore directed flow of the drifters through the rip channel, taking Eulerian measurements from the rip neck is undoubtedly important in order to analyse rip current occurrence and pulsation. The proposed data will help us to understand the life span of bathymetrically controlled rip current at different time scales, thus will add to our knowledge on defining threats and risk of rip currents at Egmond. It is recommended to perform another rip current experiment at Egmond in mid-July or August to ensure that the rip channel pattern has already been fully developed.

Shorelines from video

Shoreline position information plays an important role in two ways. First, it can provide us with intertidal bathymetry estimates which can be achieved using several methods as described in Chapter 5. As bathymetry estimation using video technique like for instance, cBathy, relies on sea surface characteristics through optical sensors, areas like intertidal bathymetry and the very shallow area near the shoreline are difficult to resolve. Having an accurate intertidal bathymetry will significantly improve overall estimates of the nearshore area. Secondly, shoreline position can be very useful to set a criterion until which pixel time stacks information is useful in providing wave motion signals, thus will help to reduce bad estimates near the shoreline. The detection method itself can be extended not only to differentiate between the sea water and the dry beach, but also to define cluster of objects appear on the beach for other kind of research and application (e.g. see Hoonhout *et al.*, 2014). Having such an algorithm that can detect shoreline or objects on the beach from video data with human un-supervised process would significantly add to not only the state of the nearshore bathymetry estimation technique from optical remote sensing, but also for coastal studies in broader topics.

cBathy further validation and application for coastal morphodynamic studies

Video image methods like cBathy and Beach Wizard have shown a very good skill in predicting nearshore bathymetry. Specifically for cBathy, the method allows us to be able to produce daily estimates of nearshore bathymetry with fully automatic process. This shows a great potential of application to be used in the operational forecasting system. During this study, only one ground truth bathymetry is available to validate the cBathy results. In the future, having a higher temporal resolution of bathymetry is recommended to evaluate cBathy skill in a more comprehensive way. In addition, the availability of this kind of data will also open the door for researchers to investigate the potential use of cBathy for coastal morphodynamic studies. Combining cBathy with process-based sediment transport model may also be studied in order to provide forecasts of bathymetry.

While cBathy has shown a very good skill in predicting nearshore features and offshore bar-trough elevation and location, problematic areas can be found with low signals, which is usually located around the offshore edge or lateral edge of the domain. In these areas, cBathy returns poor estimates, but with high confidence (low error). Low signals, can be due to the pixel quality itself that exhibits low quality, or optical disturbance like for instance sun glare, raindrops, or passing vessels. Ideally, cBathy should be able to mitigate low signal, as this kind of signal normally will also exhibit low coherence and irregular phase structure. However, there might be a situation where these faulty signals exhibit good coherence, but not within the appropriate frequency band of deep water wave motion. In this case, cBathy may continue the process using this faulty signal and return poor estimates with high confidence. Here, an investigation is recommended to analyse how the cBathy results can be different when coherence analysis (takes place in Step 1 of cBathy, Chapter 5) to be performed not based on individual tile but based on the overall domain. Once dominant frequencies are obtained based on the overall domain, matching modelled-phase-structure with observed ones then can be performed per tile, as in default cBathy. This is to avoid big difference in extracted dominant frequency over a relatively short distance (e.g. between adjacent tiles). However, it is noted that this may require a very large computational memory during the process.

Uncertainty analysis

The main prediction tool, CoSMoS, comprises many models, with different spatial scales and physics to resolve. The application of this model system is not limited to swimmer safety application; instead it was initially developed for the storm impact application. By studying uncertainty analysis of this model system, knowledge on the statistical confidence of the model results will add to our knowledge on mitigation and preparation for coastal hazards in general. This will provide decision makers and coastal manager information needed in order to examine alternatives and to undertake appropriate measures and decisions.

References

Aagaard, T., Greenwood, B. and Nielsen, J., 1997. Mean currents and sediment transport in a rip channel. Marine Geology, 140(1-2): 25-45.

Aarninkhof, S.G.J., Ruessink, B.G. and Roelvink, J.A., 2005. Nearshore subtidal bathymetry from time-exposure video images. Journal of Geophysical Research: Oceans, 110(C6): C06011.

Aarninkhof, S.G.J., Turner, I.L., Dronkers, T.D.T., Caljouw, M. and Nipius, L., 2003. A video-based technique for mapping intertidal beach bathymetry. Coastal Engineering, 49(4): 275-289.

Adler-Golden, S.M., Acharya, P., Berk, A., Matthew, M.W. and Gorodetzky, D., 2005. Remote bathymetry of the littoral zone from aviris, lash, and quickbird imagery. IEEE T. Geoscience and Remote Sensing, 43(2): 337-347.

Alvarez-Ellacuria, A. et al., 2009. An alert system for beach hazard management in the balearic islands. Coastal Management, 37(6): 569-584.

Alvarez-Ellacuria, A. et al., 2010. A nearshore wave and current operational forecasting system. Journal of Coastal Research, 26(3): 503-509.

Austin, M. et al., 2010. Temporal observations of rip current circulation on a macro-tidal beach. Continental Shelf Research, 30(9): 1149-1165.

Austin, M.J., Masselink, G., Scott, T.M. and Russell, P.E., 2014. Water-level controls on macro-tidal rip currents. Continental Shelf Research, 75(0): 28-40.

Austin, M.J., Scott, T.M., Russell, P.E. and Masselink, G., 2013. Rip current prediction: Development, validation, and evaluation of an operational tool. Journal of Coastal Research, 29(2): 283-300.

Baart, F. et al., 2009. Real-time forecasting of morphological storm impacts: A case study in the netherlands. Journal of Coastal Research, Special Issue, 56.

Baldock, T.E., Holmes, P., Bunker, S. and Van Weert, P., 1998. Cross-shore hydrodynamics within an unsaturated surf zone. Coastal Engineering, 34(3–4): 173-196.

Barnard, P. et al., 2014. Development of the coastal storm modeling system (cosmos) for predicting the impact of storms on high-energy, active-margin coasts. Natural Hazards, 74(2): 1095-1125.

Battjes, J.A. and Janssen, J.P.F.M., 1978. Energy loss and set-up due to breaking of random waves. 16th Int. Conf. Coastal Engineering, ASCE, p.^pp. 569-587.

Behrens, A. and Günther, H., 2009. Operational wave prediction of extreme storms in northern europe. Natural Hazards, 49(2): 387-399.

Bell, P.S., 1999. Shallow water bathymetry derived from an analysis of x-band marine radar images of waves. Coastal Engineering, 37(3–4): 513-527.

Benny, A.H. and Dawson, G.J., 1983. Satelite imagery as an aid to bathymetric charting in the red sea. The Cartographic Journal, 20(Number 1): 5-16.

Berens, P., 2009. Circstat: A matlab toolbox for circular statistics. Journal of Statistical Software, 31(10): 1-21.

Bertotti, L., Canestrelli, P., Cavaleri, L., Pastore, F. and Zampato, L., 2011. The henetus wave forecast system in the adriatic sea. Nat. Hazards Earth Syst. Sci., 11(11): 2965-2979.

Bertotti, L. and Cavaleri, L., 2009. Wind and wave predictions in the adriatic sea. Journal of Marine Systems, 78(Supplement 1): S227-S234.

Bierwirth, P.N., LEE, T.J. and BURNE, R.V., 1993. Shallow sea-floor reflectance and water depth derived by unmixing multispectral imagery. 59(3).

Bogle, J.A., Bryan, K.R., Black, K.P., Hume, T.M. and Healy, T.R., 2000. Video observations of rip formation and evolution. Journal of Coastal Research, SI 34: 117-127.

Booij, N., Ris, R.C. and Holthuijsen, L.H., 1999. A third-generation wave model for coastal regions 1. Model description and validation. J. Geophys. Res., 104(C4): 7649-7666.

Bowen, A.J. and Inman, D.L., 1969. Rip currents 2. Laboratory and field observations. J. Geophys. Res., 74(23): 5479-5490.

Brander, R.W. and Short, A.W., 2001. Flow kinematics of low-energy rip currents system. Journal of Coastal Research, 17(2): 468-481.

Brown, J.M., Souza, A.J. and Wolf, J., 2010. An 11-year validation of wave-surge modelling in the irish sea, using a nested polcoms–wam modelling system. Ocean Modelling, 33(1–2): 118-128.

Brown, J.M. and Wolf, J., 2009. Coupled wave and surge modelling for the eastern irish sea and implications for model wind-stress. Continental Shelf Research, 29(10): 1329-1342.

Bruijn, J.d., 2005. Rip currents, morphologically important and a hazard to swimmers, Utrecht University, Utrecht, The Netherlands.

Bruneau, N. et al., 2009. Field observations of an evolving rip current on a meso-macrotidal well-developed inner bar and rip morphology. Continental Shelf Research, 29(14): 1650-1662.

Caballeria, M., Coco, G., Falques, A. and Huntley, D.A., 2002. Self-organization mechanisms for the formation of nearshore crescentic and transverse sand bars. Journal of Fluid Mechanics, 465: 379-410.

Castelle, B. et al., 2014. Rip currents and circulation on a high-energy low-tide-terraced beach (grand popo, benin, west africa). Journal of Coastal Research(Special Issue 66).

Cavaleri, L. and Bertotti, L., 2004. Accuracy of the modelled wind and wave fields in enclosed seas. Tellus, Series A: Dynamic Meteorology and Oceanography, 56(2): 167-175.

Cavaleri, L. and Bertotti, L., 2006. The improvement of modelled wind and wave fields with increasing resolution. Ocean Engineering, 33(5–6): 553-565.

Chen, Q., Dalrymple, R.A., Kirby, J.T., Kennedy, A.B. and Haller, M.C., 1999. Boussinesq modeling of a rip current system. J. Geophys. Res., 104(C9): 20617-20637.

Chen, Q., Kirby, J., Dalrymple, R., Kennedy, A. and Chawla, A., 2000. Bousinesq modeling of wave transformation, breaking and runup: Ii. Two horizontal dimensions. Journal of Waterway, Port, Coastal and Ocean Engineering, 126: 48-56.

Cherneva, Z. et al., 2008. Validation of the wamc4 wave model for the black sea. Coastal Engineering, 55(11): 881-893.

Cornillon, P., Gallagher, J. and Sgouros, T., 2003. Opendap: Accessing data in a distributed, heterogeneous environment. Data Science Journal, 2: 164-174.

Dalrymple, R., Birkemeier, W.A. and Eubanks, R.A., 1977. Wave-induced circulation in shallow basins. Journal of the Waterway Port Coastal and Ocean Division, 103(1): 117-135.

Dalrymple, R.A. and Lozano, C.J., 1978. Wave-current interaction models for rip currents. J. Geophys. Res., 83(C12): 6063-6071.

De Kleermaeker, S., Verlaan, M., Kroos, J. and Zijl, F., 2012. A new coastal flood forecasting system for the netherlands. Hydro12 - Taking care of the sea, Rotterdam, p.^pp.

Dean, R.G. and Dalrymple, R., 1991. Water wave mechanics for engineers and scientists. World Science River Edge NJ.

Drønen, N., Karunarathna, H., Fredsøe, J., Mutlu Sumer, B. and Deigaard, R., 2002. An experimental study of rip channel flow. Coastal Engineering, 45(3–4): 223-238.

Dugan, J.P., Piotrowski, C.C. and Williams, J.Z., 2001. Water depth and surface current retrievals from airborne optical measurements of surface gravity wave dispersion. J. Geophys. Res., 106(C8): 16903-16915.

Egbert, G.D. and Erofeeva, S.Y., 2002. Efficient inverse modeling of barotropic ocean tides. Journal of Atmospheric and Oceanic Technology, 19(2): 183-204.

Engle, J., MacMahan, J.H., Thieke, R.J., Hanes, D.M. and Dean, R.G., 2002. Formulation of a rip current predictive index using rescue data. National Conference on Beach Preservation Technology, Biloxi, MS, p.^pp. 23-25.

Ferrarin, C. et al., 2013. Tide-surge-wave modelling and forecasting in the mediterranean sea with focus on the italian coast. Ocean Modelling, 61(0): 38-48.

Fisher, N.I., 1996. Statistical analysis of circular data. Cambridge University Press.

Gallop, S.K., Bryan, K.R. and Coco, G., 2009. Video obervations of rip currents on an embayed beach. Journal of Coastal Research, SI 56.

Gerritsen, H., de Vries, J.W. and Philippart, M.E., 1995. The dutch continental shelf model. In: D. Lynch and A. Davies (Editors), In quantitative skill assesment for coastal ocean. American Geophysical Union, Washington DC, pp. 425-467.

Greidanus, H., 1997. The use of radar for bathymetry in shallow seas. Hydrographic Journal, 83: 13-18.

Gunther, H., 2002. Wam cycle 4.5., Institute for Coastal Research, GKSS Research Centre, Germany.

Haas, K.A., Svendsen, I.A., Haller, M.C. and Zhao, Q., 2003. Quasi-three-dimensional modeling of rip current systems. J. Geophys. Res., 108(C7): 3217.

Haller, M.C., Dalrymple, R.A. and Svendsen, I.A., 2002. Experimental study of nearshore dynamics on a barred beach with rip channels. J. Geophys. Res., 107(C6): 3061.

Hanson, J.L., Tracy, B.A., Tolman, H.L. and Scott, R.D., 2009. Pacific hindcast performance of three numerical wave models. Journal of Atmospheric and Oceanic Technology, 26(8): 1614-1633.

Hasselman, K., Barnett, T.P., Bouws, E., Carlson, D.E. and Hasselmann, P., 1973. Measurements of wind-wave growth and swell decay during the joint north sea wave project (jonswap). Deutsche Hydrographische Zeitschrift, 8(12).

Heemink, A.W. and Kloosterhuis, H., 1990. Data assimilation for non-linear tidal models. International Journal for Numerical Methods in Fluids, 11(8): 1097-1112.

Holland, T. and Holman, R., 1997. Video estimation of foreshore topography using trinocular stereo. Journal of Coastal Research, 13(1): 81-87.

Holman, R., Plant, N. and Holland, T., 2013. Cbathy: A robust algorithm for estimating nearshore bathymetry. Journal of Geophysical Research: Oceans, 118(5): 2595-2609.

Holman, R.A. and Stanley, J., 2007. The history and technical capabilities of argus. Coastal Engineering, 54(6-7): 477-491.

Hoonhout, B., Baart, F. and Van Thiel de Vries, J.S.M., 2014. Intertidal beach classification in infrared images. In: A.N. Green and J.A.G. Cooper (Editors), International Coastal Symposium. Journal of Coastal Research, Durban, pp. 657-662.

Irish, J.L. and Lillycrop, W.J., 1999. Scanning laser mapping of the coastal zone: The shoals system. ISPRS Journal of Photogrammetry and Remote Sensing, 54(2–3): 123-129.

Janssen, P.A.E.M., Hansen, B. and Bidlot, J.-R., 1997. Verification of the ecmwf wave forecasting system against buoy and altimeter data. Weather and Forecasting, 12(4): 763-784.

Johnson, D. and Pattiaratchi, C., 2004. Transient rip currents and nearshore circulation on a swell-dominated beach. Journal of Geophysical Research: Oceans, 109(C2): C02026.

Kalman, R.E., 1960. A new approach to linear filtering and prediction problems. Transactions of the ASME – Journal of Basic Engineering(82 (Series D)): 35-45.

Kennedy, A., Chen, Q., Kirby, J. and Dalrymple, R., 2000. Boussinesq modeling of wave transformation, breaking and runup: I. One dimension. Journal of Waterway, Port, Coastal and Ocean Engineering, 126: 39-47.

Kim, I.H., Kim, I.C. and Lee, J.L., 2011. Rip current prediction system combined with a morphological change model. Journal of Coastal Research(SI 64): 547-551.

Kingston, K.S., 2003. Applications of complex adaptive systems, approaches to coastal systems, University of Plymouth, Plymouth, UK, 106 pp.

Kohut, J. et al., 2008. Surface current and wave validation of a nested regional hf radar network in the mid-atlantic bight. Proceeding of the IEEE - Conference on Current Measurement Technology, Charleston, SC, p.^pp. 203 - 207.

Kuik, A.J., van Vledder, G.P. and Holthuijsen, L.H., 1988. A method for the routine analysis of pitch-and-roll buoy wave data. Journal of Physical Oceanography, 18(7): 1020-1034.

Larsen, J., Mohn, C. and Timmermann, K., 2013. A novel model approach to bridge the gap between box models and classic 3d models in estuarine systems. Ecological Modelling, 266(0): 19-29.

Lascody, R.L., 1998. East central florida rip current program, National Weather Digest, pp. 25-30.

Lee, Z., Carder, K.L., Chen, R.F. and Peacock, T.G., 2001. Properties of the water column and bottom derived from airborne visible infrared imaging spectrometer (aviris) data. J. Geophys. Res., 106(C6): 11639-11651.

Lesser, G.R., Roelvink, J.A., van Kester, J.A.T.M. and Stelling, G.S., 2004. Development and validation of a three-dimensional morphological model. Coastal Engineering, 51(8–9): 883-915.

Leu, L.G., Kuo, Y.Y. and Lui, C.T., 1999. Coastal bathymetry from the wave spectrum of spot images. Coastal Engineering Journal, 41: 21-41.

Lippmann, T.C. and Holman, R.A., 1989. Quantification of sand bar morphology: A video technique based on wave dissipation. J. Geophys. Res., 94(C1): 995-1011.

Long, J.W. and Özkan-Haller, H.T., 2005. Offshore controls on nearshore rip currents. J. Geophys. Res., 110(C12): C12007.

Longuet-Higgins, M.S. and Stewart, R.W., 1962. Radiation stress and mass transport in gravity waves with application to surf beats. Journal of Fluid Mechanics, 13: 481-504.

Longuet-Higgins, M.S. and Stewart, R.W., 1963. A note on wave set-up. Journal of Marine Research, 21: 4-10.

Longuet-Higgins, M.S. and Stewart, R.W., 1964. Radiation stresses in water waves; a physical discussion, with application. Deep-Sea Research, 11: 529-562.

Lushine, J.B., 1991. A study of rip currents drownings and related weather factors, National Weather Digest, pp. 13-19.

MacMahan, J. et al., 2010. Mean lagrangian flow behavior on an open coast rip-channeled beach: A new perspective. Marine Geology, 268(1–4): 1-15.

MacMahan, J., Brown, J. and Thornton, E., 2009. Low-cost handheld global positioning system for measuring surf-zone currents. Journal of Coastal Research, 25(3): 744-754.

MacMahan, J.H., Thornton, E.B. and Reniers, A.J.H.M., 2006. Rip current review. Coastal Engineering, 53(2-3): 191-208.

MacMahan, J.H., Thornton, E.B., Stanton, T.P. and Reniers, A.J.H.M., 2005. Ripex: Observations of a rip current system. Marine Geology, 218(1-4): 113-134.

Madsen, A.J. and Plant, N.G., 2001. Intertidal beach slope predictions compared to field data. Marine Geology, 173(1–4): 121-139.

Mazarakis, N., Kotroni, V., Lagouvardos, K. and Bertotti, L., 2012. High-resolution wave model validation over the greek maritime areas. Nat. Hazards Earth Syst. Sci., 12(11): 3433-3440.

McNinch, J.E., 2007. Bar and swash imaging radar (basir): A mobile x-band radar designed for mapping nearshore sand bars and swash-defined shorelines over large distance. Journal of Coastal Research, 23: 59-74.

Miloshis, M. and Stephenson, W.J., 2011. Rip current escape strategies: Lessons for swimmers and coastal rescue authorities. Natural Hazards, 59(2): 823-832.

Misra, S., Kennedy, A. and Kirby, J., 2003. An approach to determining nearshore bathymetry using remotely sensed ocean surface dynamics. Coastal Eng., 47: 265-293.

Morris, J., 2013. Estimation of the nearshore bathymetry using remote sensing techniques - combining beach wizard and cbathy, Delft University of technology, Delft.

Murray, T., Cartwright, N. and Tomlinson, R., 2013. Video-imaging of transient rip currents on the gold coast open beaches. Journal of Coastal Research(SI 65): 1809-1814.

Noda, E.K., 1974. Wave-induced nearshore circulation. Journal of Geophysical Research, 79(27): 4097-4106.

Özkan-Haller, H.T. and Kirby, J.T., 1999. Nonlinear evolution of shear instabilities of the longshore current: A comparison of observations and computations. Journal of Geophysical Research: Oceans, 104(C11): 25953-25984.

Pattiaratchi, C., Olsson, D., Hetzel, Y. and Lowe, R., 2009. Wave-driven circulation patterns in the lee of groynes. Continental Shelf Research, 29(16): 1961-1974.

Plant, N., Holland, T. and·Haller, M.C., 2008. Ocean wavenumber estimation from wave-resolving time series imagery. IEEE T. Geoscience and Remote Sensing, 46.

Plant, N.G., Aarninkhof, S.G.J., Turner, I.L. and Kingston, K.S., 2007. The performance of shoreline detection models applied to video imagery. Journal of Coastal Research: 658-670.

Plant, N.G. and Holman, R.A., 1997. Intertidal beach profile estimation using video images. Marine Geology, 140(1–2): 1-24.

Ponce de León, S. and Guedes Soares, C., 2008. Sensitivity of wave model predictions to wind fields in the western mediterranean sea. Coastal Engineering, 55(11): 920-929.

Portilla, J., Ocampo-Torres, F.J. and Monbaliu, J., 2009. Spectral partitioning and identification of wind sea and swell. Journal of Atmospheric and Oceanic Technology, 26(1): 107-122.

Putrevu, U. and Svendsen, I.A., 1999. Three-dimensional dispersion of momentum in wave-induced nearshore currents. European Journal of Mechanics - B/Fluids, 18(3): 409-427.

Ranasinghe, R., Symonds, G., Black, K. and Holman, R., 2004. Morphodynamics of intermediate beaches: A video imaging and numerical modelling study. Coastal Engineering, 51(7): 629-655.

Ranasinghe, R., Symonds, G. and Holman, R.A., 1999. Quantitative characterization of rip dynamics via video imaging. Coastal Sediments, p.^pp. 987-1002.

Reniers, A.J.H.M., MacMahan, J.H., Thornton, E.B. and Stanton, T.P., 2007. Modeling of very low frequency motions during ripex. Journal of Geophysical Research: Oceans, 112(C7): C07013.

Resio, D. and Perrie, W., 1989. Implications of an f−4 equilibrium range for wind-generated waves. Journal of Physical Oceanography, 19(2): 193-204.

Roelvink, D. et al., 2009. Modelling storm impacts on beaches, dunes and barrier islands. Coastal Engineering, 56(11-12): 1133-1152.

Roelvink, J.A. and Walstra, D.J., 2004. Keeping it simple by using complex models. In: M.S. Altinakar, S.S.Y. Wang, K.P. Holz and M. Kawahara (Editors), Advances in Hydro-Science and Engineering. University of Mississippi, Oxford, MS, pp. 1-11.

Ruessink, B.G., Miles, J.R., Feddersen, F., Guza, R.T. and Elgar, S., 2001. Modeling the alongshore current on barred beaches. Journal of Geophysical Research: Oceans, 106(C10): 22451-22463.

Sandidge, J.C. and Holyer, R.J., 1998. Coastal bathymetry from hyperspectral observations of water radiance. Remote Sensing of Environment, 65(3): 341-352.

Sasso, R., 2012. Video-based nearshore bathymetry estimation for rip current forecasting on a macrotidal beach, Delft University of Technology.

Scott, T., Russel, P., Masselink, G., Wooler, A. and HShort, A., 2007. Beach rescue statistics and their relation to nearshore morphology and hazard: A case study for southwest england. Journal of Coastal Research, SI 50.

Scott, T.R. and Mason, D.C., 2007. Data assimilation for a coastal area morphodynamic model: Morecambe bay. Coastal Engineering, 54(2): 91-109.

Sembiring, L. et al., 2014. Nearshore bathymetry from video and the application to rip current predictions for the dutch coast. In: A.N. Green and J.A.G. Cooper (Editors), International Coastal Symposium. Journal of Coastal Research, Durban, pp. 354-359.

Short, A.D., 1992. Beach systems of the central netherlands coast: Processes, morphology and structural impacts in a storm driven multi-bar system. Marine Geology, 107(1–2): 103-137.

Short, A.D., 1999. Handbook of beach and shoreface morphodynamics. John Wiley & Sons, Chichester, 392 pp.

Short, A.D., 2007. Australian rip system-friend or foe. Journal of Coastal Research, SI 50.

Sonu, C.J., 1972. Field observation of nearshore circulation and meandering currents. J. Geophys. Res., 77(18): 3232-3247.

Stockdon, H.F. and Holman, R.A., 2000. Estimation of wave phase speed and nearshore bathymetry from video imagery. J. Geophys. Res., 105(C9): 22015-22033.

Svendsen, I.A., Haas, K.A. and Zhao, Q., 2000. Analysis of rip current systems. Coastal Engineering 2000, p.^pp.

Swinkels, C., 2011. Swimmer safety and rip currents indicator, Deltares, Delft.

Tolman, H.L., 2009. User manual and system documentation of wavewatch-iii version 3.14.

Turner, I.L. et al., 2001. Comparison of observed and predicted coastline changes at the gold coast artificial (surfing) reef, sydney, australia. International Conference on Coastal Engineering, Sydney, Australia, p.^pp.

Unden, P. et al., 2002. Hirlam-5 scientific documentation.

Uunk, L., Wijnberg, K.M. and Morelissen, R., 2010. Automated mapping of the intertidal beach bathymetry from video images. Coastal Engineering, 57(4): 461-469.

van der Westhuysen, A.J., Zijlema, M. and Battjes, J.A., 2007. Nonlinear saturation-based whitecapping dissipation in swan for deep and shallow water. Coastal Engineering, 54(2): 151-170.

van Dongeren, A. et al., 2008. Beach wizard: Nearshore bathymetry estimation through assimilation of model computations and remote observations. Coastal Engineering, 55(12): 1016-1027.

Van Dongeren, A., Reniers, A., Battjes, J. and Svendsen, I., 2003. Numerical modeling of infragravity wave response during delilah. Journal of Geophysical Research: Oceans, 108(C9): 3288.

Van Dongeren, A., Sancho, F.E., Svendsen, I.A. and Putrevu, U., 1994. Shorecirc: A quasi 3-d nearshore model. 24th International Conference on Coastal Engineering, Kobe, Japan, p.^pp. 2741-2754.

Van Dongeren, A. et al., 2013. Rip current predictions through model data assimilation on two distinct beaches. Coastal Dynamics, Bordeaux, France, p.^pp. 1775-1786.

Van Ormondt, M. et al., 2012. Simulating storm impacts and coastal flooding along the netherlands coast. Flood Risk 2012, Rotterdam, The Netherlands, p.^pp. 28-29.

Van Son, S.T.J., Lindenbergh, R.C., De Schipper, M.A., De Vries, S. and Duijnmater, K., 2009. Using a personal watercraft for monitoring bathymetric changes at storm scale. In: A. Price, J. Raubenheimer, M. Gilissen, D. Sinclair and S. Smith (Editors), International Hydrographic Conference Cape Town, South Africa, pp. 1-11.

Verlaan, M., Zijderveld, A., de Vries, H. and Kroos, J., 2005. Operational storm surge forecasting in the netherlands: Developments in the last decade. Philosophical Transactions of the Royal Society A: Mathematical, Physical and Engineering Sciences, 363(1831): 1441-1453.

Vittori, G., DE SWART, H.E. and BLONDEAUX, P., 1999. Crescentic bedforms in the nearshore region. Journal of Fluid Mechanics, 381: 271-303.

Walstra, D.J.R., Reniers, A.J.H.M., Ranasinghe, R., Roelvink, J.A. and Ruessink, B.G., 2012. On bar growth and decay during interannual net offshore migration. Coastal Engineering, 60(0): 190-200.

Wei, G., Kirby, J.T., Grilli, S.T. and Subramanya, R., 1995. A fully nonlinear boussinesq model for surface waves. Part 1. Highly nonlinear unsteady waves. Journal of Fluid Mechanics, 294(0): 71-92.

Wenneker, I. and Smale, A., 2013. Measurement of 2d wave spectra during a storm in a tidal inlet, Coastal Dynamics, France.

Werner, M. et al., 2013. The delft-fews flow forecasting system. Environmental Modelling & Software, 40(0): 65-77.

Wiersma, J. and Van Alphen, J., 1987. In: P.L. De Boer, A. Van Gelder and S.D. Nio (Editors), Tide-Influenced Sedimentary Environments and Facies. D. Reidel, Utrecht, pp. 101-111.

Wijnberg, K.M., 2002. Environmental controls on decadal morphologic behaviour of the holland coast. Marine Geology, 189(3-4): 227-247.

Wilson, G.W., Özkan-Haller, H.T. and Holman, R.A., 2010. Data assimilation and bathymetric inversion in a two-dimensional horizontal surf zone model. J. Geophys. Res., 115(C12): C12057.

Winter, G., van Dongeren, A.R., de Schipper, M.A. and van Thiel de Vries, J.S.M., 2014. Rip currents under obliquely incident wind waves and tidal longshore currents. Coastal Engineering, 89(0): 106-119.

Wolf, J., 2008. Coupled wave and surge modelling and implications for coastal flooding. Adv. Geosci., 17: 19-22.

Wright, L.D. and Short, A.D., 1984. Morphodynamic variability of surf zones and beaches: A synthesis. Marine Geology, 56(1-4): 93-118.

Wu, C. and Liu, P., 1985. Finite element modeling of nonlinear coastal currents. Journal of Waterway, Port, Coastal, and Ocean Engineering, 111(2): 417-432.

Acknowledgement

This study would not have been possible without financial support from Deltares. For that, I would like to express my special thanks and appreciation to Prof. Dano Roelvink and Dr. Ap van Dongeren, for their efforts on creating a PhD research project at Deltares for me. I am proud to be a PhD student under their supervision. I would like to express my thanks for all the discussions and inputs they provided during my study. I really feel that I am in the right hands of experts. I also thank my PhD Committee member for the constructive comments to the thesis.

I would like to thank Maarten van Ormondt, for all the support he provided when I was working with the CoSMoS. I thank the Argus Team of Deltares: Christophe, Giorgio, Irv, Bas, and Robin. Thanks for the supports you provided regarding the Argus camera system of Deltares.

The cBathy Chapter in this thesis would not have been possible without the support of Prof. Rob Holman from the Oregon State University. Thank you for making the cBathy code available for me through Deltares.

There are so many people I don't know by person, but contribute and helped a lot so I can finish this thesis. These are specially people in the discussion forum on the internet, where Matlab tips, tricks, and files are shared, I would like to express my big thanks.

I had an opportunity to work with good MSc students: Gundula Winter (TU Delft), Roberto Sasso (CoMEM TU Delft), Jamie Morris (TU Delft), and Dimitris Chatzistratis (Utrecht Univ). I thank them for the good time and discussions, which contribute to this thesis as well. I also thank the many volunteers during the SEAREX field campaign, under coordination of Gundula, which contribute very much to the Chapter 4 of this thesis.

I would like to thank Johan Reyns and MSc students of IHE batch 2012-2014. Thanks for helping me out during my field experiment at Egmond aan Zee.

I would like to thank Arnold van Rooijen for his long distance assistance when I was in Indonesia, when my laptop was stolen at the Zoutstate office. I wish him lots of luck for his coming PhD study in Australia.

I thank my former XBeach-room-mate at Deltares. Robert McCall and Jaap van Thiel. I am not really into the Johnny Cash's music, but some of the songs are actually quite enjoyable. For me, it's an honour to be one of the occupants of that famous room.

I want to thank my coastal colleagues at IHE: Trang, Guo, Fernanda, Duoc, Liqin, Abdi, and Rita. It always feels good every time we can share our experiences and hurdles. I also thank Ali Dastgheib for being not only a nice and supportive lecturer, but also a cool friend.

I am very grateful that living in Delft far from home, can sometime be as homey as home. I express my thanks to Isnaeni and family, Yuli (for picking up the 'penganten baru' at Schipol), Shah, Fiona, Sony, Pak Anuar, Selvi, Clara, Tarn and Quan Pan. I specially thank Marwan and Ririn for taking care of my son when my daughter was born. I thank Dr. Suryadi for constant effort in motivating, and for being always available for any kind of discussions.

I would like to remember late Kak Ade, who assisted me and my wife a lot during the born of my son. Her memory will always be with our family.

I thank the Djayadisastra family in Bandung, Mang Dede and the family, Kang Soni and the family. Thank you for preparing and providing so many helps so Eyang was able to visit Delft for more than a month while I was so busy with writing the thesis. Thanks for Tante Rezi for always caring to Eyang, especially during the trip.

It was a hard time for me and my wife when our beloved Bapak, Rachmat Natawidjaja, passed away, and we could not be on his side in Bandung during his last time. We believe he is in a much better place now.

Finally, I am grateful to have my parents, who always supportive and knows how to cheer me up. Thanks that you always proud of me no matter what. I thank my older sister, Sion, also Bang Bergi, and lovely Audra, and my little sister Ima, for all your supports during my study. Thank you for taking care of Mamak and Bapak with full of love while I was far away from home.

Doing a PhD and having a two baby-born at the same time, far from home, is definitely a special achievement in my life. Dewi Kurnia has been unconditionally supportive in taking care of me and our two lovely kids. I am so grateful to have her in my life, and looking forward to our next journey.

Delft, 2015

Leo Eliasta Sembiring.

List of Publications

First Author

Sembiring, L., Van Dongeren, A., Winter, G., Van Ormondt, M., Briere, C., Roelvink, D., 2014. Nearshore bathymetry from video and the application to rip current predictions for The Dutch Coast. In: Green, A.N., and Cooper, J.A.G. (eds), *Proceedings 13th International Coastal Symposium* (Durban, South Africa), *Journal of Coastal Research*, Special Issue No.70, pp. 354-359, ISSN 0749-0208.

Sembiring, L., van Ormondt, M., van Dongeren, A. and Roelvink, D., 2015. A validation of an operational wave and surge prediction system for the Dutch coast. *Nat. Hazards Earth Syst. Sci.*, 15(6): 1231-1242.

Sembiring, L., Van Dongeren, A., Roelvink, D., Winter, G. Dynamic modelling of rip current for swimmer safety on wind-sea meso-tidal beach. (*In press*), *Journal of Coastal Research*.

Co-Author

Van Ormondt, M., Van Dongeren, A., Briere, C., Sembiring, L., Winter, G., Lescinski, J., Swinkels, C. 2012. Simulating storm impacts and coastal flooding along the Netherlands coast. *Flood Risk 2012*, Rotterdam, The Netherlands, pp. 28-29.

Van Dongeren, A., Van Ormondt, M., Sembiring, L., Sasso, R., Austin, M., Briere, C., Swinkels, C., Roelvink, D., Van Thiel De Vries, J. 2013. Rip current predictions through model data assimilation on two distinct beaches. *Coastal Dynamics*, Bordeaux, France

Conference talks:

- Rip current prediction system. Oral presentation at PhD Symposium UNESCO-IHE, Delft, 2012.
- Estimation of remote sensed bathymetries for forecasting of rip currents. Oral presentation at Rip Current Symposium, Sydney, 2012
- Nearshore operational for rip current predictions. Oral presentation at Ocean Science Session, AGU Fall Meeting, San Francisco, 2012.
- Nearshore bathymetry from video and rip current predictions. Oral presentation at PhD Symposium UNESCO-IHE Delft, 2013
- Rip current prediction through numerical model and bathymetry from video. Oral presentation at NCK Days 2014, Delft.

About the author

Leo Sembiring was born on 2 August 1980 in Kabanjahe, Indonesia. He finished his high school in Medan, North Sumatera, where his parents reside, and continue his university study at Civil Engineering Department of Bandung Institute of Technology (ITB), in West Java. He obtained his B.Sc in 2004 after writing a thesis on developing peak ground acceleration map due to tectonic earthquake for Sulawesi Island. Shortly after his graduation, he joined the Research Centre for Water Resources, Ministry of Public Works, also in Bandung, as a Research Scientist. Within this ministry, he works mainly on geotechnical engineering aspects of hydraulic structures. He was involved in geotechnical field surveys and investigations as well as data analysis and designs. Moreover, he was also involved in some researches on the safety of several large dams throughout Indonesia.

In 2008, he started his graduate program in Coastal Engineering and Port Development program in UNESCO-IHE, Delft, during which he was awarded the "Studeren in Nederland" (StuNed) scholarship from the Dutch government, which covers fully his living cost and the tuition fee for his study. He obtained his M.Sc degree in 2010 (with distinction). He carried out his master research at Deltares, working on the validation of wave and hydrodynamic model for The Dutch Coast (SWAN and DELFT3D) and application of data assimilation model Beach Wizard-dissipation maps in updating nearshore bathymetry.

In June 2011, he started his PhD work on developing an operational forecasting system of rip currents for Egmond aan Zee beach. During this period, he spent most of his time at Deltares and occasionally at UNESCO-IHE. During his research, he was involved in SEAREX field campaign at Egmond aan Zee (August 2011), and responsible to organize and conduct his own field campaign on June 2013, during which he was assisted by MSc students from UNESCO-IHE.

Leo is married to Dewi Kurnia. They have a son, Sagara (3), and a daughter Aruna (10 months).

T - #0430 - 101024 - C152 - 244/170/8 - PB - 9781138029408 - Gloss Lamination